SCIENCE

FOR THOSE WHO ESCAPE FROM SCIENCE

By Navid Farrokhi

First edition published in 2021
© Copyright 2022
Navid Farrokhi

The right of Navid Farrokhi to be identified as the author of this work has been asserted by him in accordance with the Copyright, Designs and Patents Act 1998.

All rights reserved. No reproduction, copy or transmission of this publication may be made without express prior written permission. No paragraph of this publication may be reproduced, copied or transmitted except with express prior written permission or in accordance with the provisions of the Copyright Act 1956 (as amended). Any person who commits any unauthorised act in relation to this publication may be liable to criminal prosecution and civil claims for damage.

All characters appearing in this work are fictitious. Any resemblance to real persons, living or dead, is purely coincidental. The opinions expressed herein are those of the author and not of MX Publishing.

Paperback ISBN 978-1-80424-023-6
ePub ISBN 978-1-80424-024-3
PDF ISBN 978-1-80424-025-0

Published by MX Publishing
335 Princess Park Manor, Royal Drive,
London, N11 3GX
www.mxpublishing.com

Cover design by Brian Belanger

Table of Contents

Introduction	4
Beyond the Earth	6
Two-legged human	32
Terrestrial living	55
Emperor of the Sea	79
Shah of Heaven	89
Small Animals	102
Mother Earth	115
Inventions	140
Pictures	153

Introduction

Science helps us to discover the universe and the world around us and aids us in understanding strange and wonderful things. The greatest human progresses and advances came true when man doubted his knowledge and sought to discover new challenges. Unfortunately, despite our advances, many people are still unaware of many facts about ourselves and that there is a lot of things that we need to know about.

Science has interesting aspects, but in many cases, its mesmerizing aspects are hidden in the form of textbooks and educational process. For those interested in the science professionally, more or less, good works have been written, but it is rare to find a book whose target group is people who try to escape from science! Therefore, the purpose of the present book should be considered an attempt to fill this gap.

In the vast ocean of human knowledge, the breadth of the various sciences is so great that this work is only a drop against it. The book consists of eight main chapters in five different areas: the first chapter deals with facts "beyond the Earth," the second chapter deals with biological knowledge related to "human beings," the third, fourth, fifth and sixth chapters deal with facts about other living things on the planet Earth (birds, aquatic animals, and terrestrial beings). Chapter 7 is about the Earth itself and its nature, and finally, chapter 8 will summarize the seventeen great inventions of history that have changed human life.

Do you know which organ of the human body has five hundred different functions? Do you know how many species of animals become extinct each year? Do you know which animal fingerprint is indistinguishable from a human fingerprint? Do you know which tree can live a thousand years? Did you know that the radio was not actually invented by the Italian Marconi?

No? So why are you procrastinating?! Turn the page and start reading! Hope you enjoy!

Navid Farrokhi

Beyond the Earth

- All the messages and signals that have been sent by humans to communicate with extraterrestrial life haven't gone more than 200 light-years away. The diameter of the Milky Way galaxy alone is 120,000 light-years. That means even 99% of our own galaxy has not yet received our message!
- Frank Drake, a retired professor of astronomy and astrophysics developed an equation that bears his name. According to Drake's complex equation, there are thousands of intelligent civilizations in the Milky Way galaxy alone.
- By calculating approximately all the planets in the universe, scientists estimate that the probability that human beings are alone and there are no other intelligent beings in the universe is one in 100 billion!

- If extraterrestrial beings from a planet four billion light-years away receive a message of Earth, it would be useless to respond, because by then the sun in the solar system would be extinguished and the situation on Earth would have completely changed!
- According to some evolutionary biologists, babies born outside the Earth's atmosphere will have larger heads depending on gravity! They believe if a human gives birth to a child somewhere other than Earth, that baby will not be similar to children of Earth.
- If we were born on a planet with half of the Earth's gravity, we would have evolved to be as tall and thin as the creatures in the movie *Avatar*.

- If we consider the age of the Earth equal to the life of an eighty-year-old creature, humanity is only eight hours old, and it's just two minutes since the Renaissance, and only fifteen seconds since we first left Earth's atmosphere!
- There is generally no hypothetical line for the starting point of space and exit from the Earth's atmosphere; however, the Karman line, located 100 km above sea level, is typically considered the starting point for space.
- The sun is the main source of light, energy, heat and life on Earth, and its dimensions are so large that it can fit one million Earths inside it! In the ancient Persian, sun was called by names such as Khor, Hoor, Mehr, Rooz.

- The temperature of the sun varies in different points from about 15 million degrees Celsius to only around 5,500 degrees.
- Mercury is the closest planet to the Sun; in Persian it is known as "Tir" which means "Shot" because it is the "fastest" planet in the solar system. Mercury orbits the Sun at a speed of approximately 48 kilometers per second, once every 88 days.
- Every day on Mercury is equivalent to 59 days on Earth. Due to the proximity of this planet to sun, if you are in front of the sun, it will cook you quickly at a temperature of 465 degrees. Ironically, if you are on the dark side, you will freeze due to the -148 degree Celsius temperature in a short time!
- Mariner 10 spacecraft passed close to Mercury in 1974 and took about 3,000 photos in one year. Mariner 10 is still orbiting the sun.
- For a person living on Venus, a day on this planet takes more than a year! Although it sounds contradictory, it is true. Every time a planet turns on its axis, it counts as a day, and every time it revolves around the sun, it counts as a year. It takes 243 days (Earth time) for the planet Venus to cycle around itself once, while it take 225 days to revolves around Sun!
- Venus is the second closest planet to the Sun in the solar system. Venus is the brightest object in the Earth's sky after the Moon and is called

"Nahid" in Persian language which is taken from the name "Anahita", an ancient Iranian goddess.
- Venus is similar to Earth in many ways, such as size, mass, gravity, and structural composition, which is why many have called it "Earth's sister".

- In the midst of the Cold War, the Soviet Union sent the Venera-13 spacecraft to Venus in 1981. The spacecraft successfully landed on the planet, but due to little knowledge of Venus' atmosphere, Venera-13 melted after 127 minutes. Venus is the hottest planet in the solar system.
- The air pressure of Venus is more than 90 times higher that of Earth. This means that when you walk on the planet Venus, you feel like you are walking at a depth of 1000 meters below the sea.
- The Persian name of the planet Mars is "Bahram" which is the god of war in ancient

Persian. Mars was also known as the god of war and bloodshed because of its red color.
- The planet's soil is alkaline and contains substances such as magnesium, sodium, potassium and chlorine, which are essential for life.
- Absorption of oxygen in one breath on Earth for humans is equivalent to 14,000 breaths on Mars!
- Early probes that landed on Mars showed that the Red Planet was a dry, desolate place away from any signs of life. But discoveries in recent decades have shown that there is ice and large amounts of nitrogen, oxygen and carbon (essential for life) in the polar regions of Mars. In the left and right images, you can see dried rivers on Mars and Earth.

Dried River on Mars   Dried River on Earth

- The average temperature of Mars is -63 degrees Celsius, its gravity is about 40% of the Earth and the deadly rays in it are 5,000 times more than Earth! However, 200,000 people signed up for a no-return trip to Mars! Of the 100 selected finalists to travel to Mars, three were Iranians.
- Mars has a mass less than Earth, and this makes a person weighing 80 kg on Mars weigh about 30 kg there!
- Time on Mars is a little faster than Earth! Unlike Newton, who saw gravity as a force, Einstein interpreted it as a kind of curvature, and according to the theory of general relativity, the Earth's greater gravity causes time to pass more slowly than on Mars.

- The highest peak ever discovered by man in the entire universe is Mount Olympus. The mountain is located on the surface of Mars and its height is about 25 km, almost three times that of Everest.

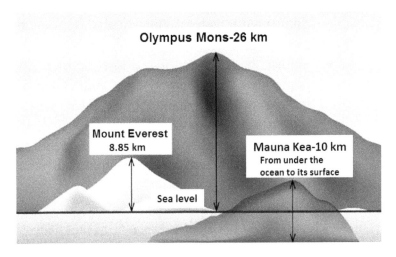

- Jupiter or Hormoz (in Persian) is the largest and oldest planet in the solar system. About 1300 planets the size of Earth can fill Jupiter! Hormoz was the life-giving god of the ancient Iranians.
- Some storms on Jupiter are larger than the entire planet Earth.
- Jupiter's gravity keeps the Earth away from asteroids and meteors and keeps the situation in the solar system stable.

- Saturn or Keivan (in Persian) looks beautiful because of the bright rings around it. Keivan means great character in ancient Persian.
- Saturn's rings are made of ice with a purity of 99.9% and fragments of toluene and silicate.
- Galileo was the first to observe the rings around Saturn with his telescope. At first he thought the planet had ears!
- Saturn has 82 moons, most among all the planets in our solar system, and Titan is the largest with a diameter of 5150 km. The interesting thing about Titan is that it has enough atmosphere for chemical reactions, as well as lakes and rivers of hydrocarbons and methane.

- Pioneer 11 was the first spacecraft to get close to Saturn and its rings in 1979, and it was the second spacecraft to pass Jupiter after Pioneer 10.

- Uranus is the coldest planet in the solar system with a temperature of -224 degrees.
- Uranus orbits the Sun once every 84 years (Earth time), so when a person is born on Earth and dies in the mid-eighties, he is just one year old on Uranus!
- Neptune was first observed by Galileo in 1613, but on that time he thought it was just another star.
- In fact, the biggest factor in discovering Neptune was the planet Uranus! By calculating the orbit of Uranus, astronomers realized that the planet's motion did not follow normal patterns, and this irregular orbit convinced them that an undiscovered planet was the cause of this disorder!
- Neptune is the only planet invisible to the naked eye.
- The only human spacecraft to visit Neptune so far was the Voyager 2 spacecraft that passed by the planet in 1989 during a large solar system tour.
- Neptune was known as farthest planet of the solar system from the discovery of Pluto in 1846 until 1930. But when Pluto was reduced to a dwarf planet in the new century, Neptune regained the title of the farthest planet!
- It takes 248 years on the Earth's calendar for Pluto to orbit the Sun once, and counting backward that number is roughly equal to the

date of American Independence! In other words, for a hypothetical inhabitant of the dwarf planet Pluto, the lifespan of the US government is about a year!

- In 2006, the International Astronomical Union declared Pluto unfit to be a planet. The reason for this was that Pluto did not satisfy their definition of a planet. According to the union's definition of a "planet", it should:
  1. Orbit the sun and have an almost spherical shape.
  2. Not be a moon of another celestial mass.
  3. Steal small objects from its surroundings.
- The decision of the International Astronomical Union on Pluto was not universally approved. So many associations protested and even a group started collecting signatures to turn Pluto back into a planet!

- Most of the images published in textbooks of the solar system show only the sun and the planets, but moons and other objects are inevitable parts of the solar system (it is also important to note that solar systems objects are not in same orbits and similar angles and in fact no two-dimensional design shows the apparent truth of the solar system).

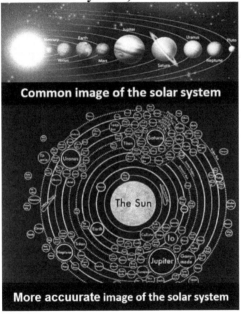

Common image of the solar system

More accuurate image of the solar system

- The Parker Solar Probe, an unmanned spacecraft launched to test the solar corona on August 12, 2018, traveling at 340,000 kilometers per hour. It is the fastest man-made device ever built. However, even with this high-speed spacecraft, it would take tens of thousands of years to reach the nearest extrasolar planet, Proxima B!

- The energy of just one of the solar flares that radiates from its surface is equivalent to the explosion of millions of 100-megaton atomic bombs. Needless to say, the two bombs that destroyed the cities of Hiroshima and Nagasaki during World War II released a total of only 35 kilotons of energy.

- The sun is also in orbit. Of course, different parts of it do not rotate at the same speed. Astronomers first discovered this from the motion of sunspots. A full cycle of the sun takes about 25 days in the center and about 36 days in the polar regions.

- If one day you were asked what color the sun is, what would you answer? You can probably say with confidence that the sun is yellow. Or maybe with a little thought, choose orange or red. But the

sun is essentially white! However, it is seen as yellow due to the passage of the Earth's atmosphere and due to the Rayleigh scattering of the atmosphere.
- If we heat only the head of a needle to the temperature of the hottest spot on the sun, all living things within a radius of 3,000 km will turn to ashes. Needless to say, the temperature of the center of the sun is about 15 million degrees Celsius.

- The sun has sound, but this sound is not heard due to the lack of particles that transmit sound waves.
- Although it is estimated that when the sun turns into a red giant (when the hydrogen fuel in its core is completely used up), it also covers the Earth's surface, but its gravity on Earth may also decrease as its mass decreases. In other

words, at the same time of the sun's destruction, the Earth is likely to move away from the it.
- According to quantum theory, there are no concepts of "present," "past" or "future"! All time situations are interdependent and all universe events are moving at the same speed!
- 99.99999999% of the space of an atom is empty! This means that the book you are reading, the cell phone you are holding, the house you live in and even yourself are almost non-existent! If the atoms did not have space, we could fit the whole globe in one baseball, while the weight of the baseball would be as heavy as the globe!
- Part of the Earth's atmosphere is made up of gases released from the craters of volcanoes. Therefore, volcanoes can be considered as one of the main actors in the existence of life on the planet Earth.
- About 65 million years ago, a meteorite-shaped asteroid almost 14 kilometers in diameter with a magnitude of 7 billion times that of an atomic bomb, hit the Yucatan Mexico region and dust from the soot covered the entire Earth's atmosphere. As a result of the collision, 75% of life species, including dinosaurs, became extinct.
- If we could look at the Earth with a telescope from a planet 70 million light-years away from Earth, we can see dinosaurs that have not yet become extinct and rule the Earth!

- Every human being has two eyes. Each eye has 130 million light-receiving cells. Each of these cells also contains 100 trillion atoms, which is more than the number of stars in the Milky Way!
- The noise that occasionally appears on television channels and the sounds we hear as channels change in an instant are due to radiation left over from the Big Bang (which occurred 13.7 billion years ago).
- The shortest march on the surface of the moon was two and a half hours and the longest was 22 hours.
- So far (January 2022), 12 people have traveled to the moon, but only 3 persons have gone to the deepest part of the ocean. We know Mars and the moon better than the depths of the Earth's oceans!
- The moon moves 4 cm away from the earth every year. If this trend continues for a very long time, it could be the first cause of human extinction, because as the moon moves away, the waters will rise and cover the whole earth.

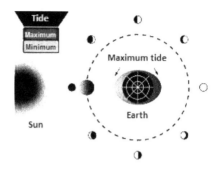

- The wind never blows on the moon, so the footprints of astronauts on the Apollo missions are still there!
- Early in Earth's existence, the moon was 100 times brighter (reflect the sun's light much more than today) since it was about 10 times closer to Earth. Also, the tide was several times higher now.
- Darkness of space does not mean the absence of light. Outer space is full of the light of stars like the sun, but since the eye cannot see the direction of light, we think it is dark. The reason why light is not seen is that there is nothing in space, not even air molecules, and there is an absolute vacuum, and because of this, light cannot be reflected, and as a result, everything looks dark and black.
- Outside the Earth's atmosphere, sound has no meaning. What causes different sounds to be heard on the Earth are the air molecules that vibrate. But there is no molecule in space that transmits sound.

- The gravitational pull of the planets Jupiter and Venus causes the Earth's orbit to change every 405,000 years. Naturally, this can change the Earth's climate patterns.
- The speed of a spacecraft leaving the Earth's atmosphere is 11.5 kilometers per second. With this speed, you can walk from Tehran to Babol (Cities in Iran) in 18 seconds.
- In the 1960s, when the Apollo mission began, the spacecraft's computers had lower processing power than today's smartphones!
- NASA astronaut Charles Moss Duke Jr. is the tenth human to land on the moon during the Apollo 16 mission. There, he posted a family photo of himself, his wife and children on the moon, with the caption: "This is the family of astronaut Charlie Duke from planet Earth who landed on the moon on April 20, 1972."
- If we accidentally fall to the Earth from space for 1000 times, we would fall 708 times in the oceans and seas, 175 times in the deserts and mountains, 116 times in the fields, and only once would we fall in the city or village!
- Archimedes claimed that the earth could be lifted with a lever. His claim is scientifically possible. If his lever is 8.4 million light-years long and he pushes it down 10,000 light-years, then the Earth will be movable (it would shake one millimeter!).

- Scientists estimate that the Earth's rotation slows by about 1.4 milliseconds per century. This means that in the time of the dinosaurs, a day was about 23 hours.
- In a 2012 study, researchers at the University of Rochester and Leiden discovered a planet called J1407b, whose rings are 200 times larger than those of Saturn. The planet is approximately 430 light-years from the solar system.

- Light travels a distance of 18 million kilometers in one minute. So in one hour, it travels about one billion and eighty million kilometers and in one day 25 trillion and 920 billion kilometers. If you multiply the number obtained by 365, the product is equal to the concept of one light year.
- The lifespan of the universe is about 14 billion years, which means that light can travel a maximum of 14 billion light-years from the beginning of universe.

- The speed of light is the highest speed we have ever seen, but it is not that fast considering other astonishing cosmic distances! To travel out of the Milky Way galaxy alone, we must travel 150,000 light-years, and to reach the nearest galaxy, Andromeda, we must travel 2.5 million light-years.
- The force of gravity between the Milky Way and the Andromeda Galaxy pulls them together, and the two galaxies are getting closer to each other!
- If we put together 25 Milky Way galaxies, we will reach the Andromeda Galaxy, and if we continue this arrangement 5500 times, we will reach the border of a world visible to humans!
- The earth revolves around the sun at a speed of 30,000 meters per second and rotates at an approximate speed of 450 meters per second. The reason that nothing like the picture below happens to the Earth and everything is not thrown out is just because of gravity!

- The Pacific Ocean has an area of 165 million square kilometers (46% of the Earth's surface water). Often the size of the ocean is not given enough attention. The Pacific Ocean is so large that from some angles outside the Earth, it seems it covers the entire surface of the planet for an outside observer.

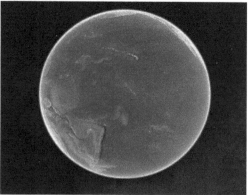

- If a person holds his breath, he can survive in space for about 30 seconds without any protection.
- There are doubts about whether Earth is our main planet. According to Darwin's theory of natural selection, creatures adapt to the nature around them, but humans are damaged by direct sunlight and suffer from back pain and chronic diseases over time due to gravity. Some researchers therefore believe that Earth is not our main planet and that our real planet should have less gravity.
- Scientists estimate that in the distant future, all matter in the universe will be converted into energy, and due to the great expansion of the universe, there will be thin, scattered dust from light forever in the universe.
- One of the efforts of space organizations is to enable human life outside the Earth's atmosphere independently of the natural resources of the planet Earth. To this end, on August 10, 2015, for the first time, NASA astronauts ate food that was completely processed in space!
- The United Nations has enacted special rules for space so that space does not become a place for battles or nuclear tests. For example, no government is allowed to bring weapons of mass destruction into orbit, and space exploration is allowed only for peaceful purposes. In the meantime, any country that

sends something into space is responsible for the damage caused by it.
- The first living thing to be sent into space was a stray dog named Laika. This unfortunate animal was selected to be sent on a space mission randomly from wandering the streets of Moscow. Laika was not very lucky and died on November 3, 1957, on the Sputnik 2 spacecraft, after 5 to 7 hours due to extreme heat and stress.

- Scientists estimate that about 1,600 trillion tons of gold are hidden in the center of the Earth. If all of the Earth's gold were spread evenly over the Earth's crust, a 46-centimeter-thick layer of gold would cover the entire earth.
- Sometime ago, the Chinese Space Agency's powerful telescope detected more than 100 fast radio waves from an unknown source 3 billion light-years away from Earth!
- Established with the participation of more than 15 countries, the International Space Station is

the most expensive man-made structure to date at a cost of 150 billion dollars. The station is about the size of an American football field.
- More than 90% of Earth's inhabitants are able to watch the International Space Station from Earth. This station looks like a star moving at high speed during the night. Just visit Heavens-Above.com to find out when the International Space Station will pass through your view.
- The Hubble Space Telescope observes a planet that looks like black asphalt 1,300 light-years away from the solar system. Instead of reflecting light like any other planet, this strange planet swallows the light! The planet is called WASP-12b, and it traps 94% of visible light inside like a black hole.
- In the image below, three slender stems of a greenish haze appear, like the small minarets of castles in science-fiction stories. Blue streams are scattered on their tips. Shining stars glow around these gas towers, leaving bright yellow patterns. This image may seem like a fantasy work to you, but these foggy castles are absolutely real! Here is an area of the Eagle Nebula called the "Pillars of Creation." This space environment is covered with interstellar gases and cosmic particles, 7,000 light-years from Earth.

- We are all actually made of stars! The main constituent of our cells is carbon, and carbon was first created in stars, so the roots of our bodies and the stars are the same!
- You are one of 7 billion people on one planet out of 8 planets, in the orbit of one star out of 400 billion stars in a galaxy out of 2000 billion galaxies in the universe!
- Stars may be considered time machines because the farther away you look, the longer you go back in time and you see the light that was emitted thousands of years ago and now it reaches my eyes and yours.
- The largest galaxy ever observed is the galaxy IC 1101. This galaxy has more than 100 trillion stars! By the way, this galaxy is sixty times the

size of the Milky Way and dates back to thirteen billion years ago, almost the beginning of the Big Bang.
- 275 million stars like our sun are born or die every day around the universe. The number of these births in one earth year reaches more than 100 billion.
- A black hole is a very dense region in space-time that attracts even light due to its strong gravitational pull.
- Scientists believe that at the center of each galaxy is a large, massive black hole that helps keep the various components of the galaxy in correct order. The supermassive black hole at the center of the Milky Way galaxy has a mass 4 million times that of the Sun.
- Black holes sometimes appear as binary forms and rotate in close orbits. Astrophysically, these black holes are the strongest source of gravitational waves in the world.

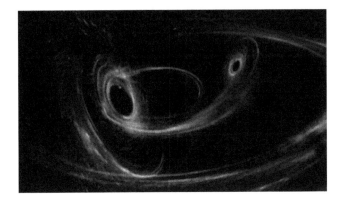

- The wonder of creation is enough that although the number of stars in the universe is estimated to be greater than the number of grains of sand on Earth, the number of atoms on one grain of sand is more than the number of stars in the universe!

Two-legged humans

- Babies have about 270 bones, which makes their skeleton more flexible than an adult. As we age, many of the bones fuse together, leaving 206 bones.
- The smallest bone in the human body is in the middle ear, which is only 2.8 mm long.
- Bones come in a variety of shapes and sizes, with more than half of the body's bones located in the arms and legs (106 bones).
- The toes are the most fragile bones in the body. Many people experience a broken toe bone once in a lifetime.
- Bone lacks a nervous system to control movement. In fact, when a person moves or moves his limbs, he commands the muscles that are attached to the bones.

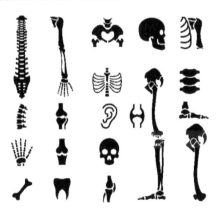

- Human hair grows faster in the beard area than anywhere else in the body. If an adult man's beard is never shaved in his lifetime, it can reach up to 90 meters!
- Human body hair grows faster than during the summer days. Hair growth rate decreases in winter and at night.
- The human body has as much hair as a chimpanzee, but most of this hair is so thin and useless that it is almost invisible to the naked eye!
- Human body hair is virtually indestructible and remains healthy even after thousands of years.
- All mammals have hair. One of the main functions of hair is to regulate heat in the body. However, it is said that compared to early humans, our hair has become thinner due to the evolution process and has lost its main role.

- Except for the palms and soles of the feet, there are fine hairs on all parts of the human body.
- Man is constantly shedding skin. Each person sheds about 600,000 particles per hour! That is about 680 grams per year. When a human being reaches the age of 70, he has had about 50 kilograms of skins!
- Human skin as a whole weighs about 3 kg.
- The intestine is an important part of the digestive system that plays a very important role in the digestion and absorption of food. The human intestine is a total of 9 meters long and is spiral shaped.
- The small intestine, about six meters long, is the largest internal organ in the human body!
- Feeling cold during the night leads to nightmares. The colder the room, the more likely you are to have a bad dream. So sleep in a warm room at night!
- Humans are taller in the early morning than in the evening. During the day and when a person is moving, the spine is compressed due to the weight of the body, and when you are asleep, It comes out of compression.
- Insomnia is 1.4 times more common among women than men.
- Randy Gardner, an American student, holds the record for the longest period of intentional awakening. He stayed awake for 264.4 hours (11 days 25 minutes).

- Research shows that 26 minutes of naps increases alertness by 54%.
- Many ancient therapies really worked! One of them is placing honey on the wound and covering it with willow bark. Honey has antiseptic properties and prevents the wound from getting worse, and the willow tree has the same pain reliever found in aspirin!
- If a wound becomes infected, a leech worm can be placed on it. This insect eats infected spots and prevents them from spreading!
- In the past, doctors used leech worms to draw human blood. Leeches are still used today because they produce chemical compounds that relieve pain and prevent blood clots.
- Rayu is a type of Chinese condiment. Pouring some dense Rayu on open wounds during surgery numbs the nerves for weeks and keeps patients from feeling pain after surgery.
- Sutures are known as a way to heal most wounds. In ancient times, Native American physicians used forest ants to suture wounds!

They held the edges of the wounds together and brought the ant to the wound to bite that part of the skin. The ant's head was then amputated at the same time, and the ant's jaws acted as sutures!
- A traditional way to treat arthritis pain is bee stings. Of course, after that you have to endure the pain of the sting!
- The human heart goes through different stages of development. Our heart in the embryo is similar to the heart of a fish. It then resembles the heart of a frog with two valves, then resembles the heart of a snake and a turtle with three valves, and finally becomes the heart of a human with four valves.

- The main function of the heart is blood circulation in the body. The human heart weighs about 300 grams.
- The human heart beats about 35 million times a year.
- Examinations of a 3,000-year-old mummy showed he had a heart problem.
- On average, women's heart beats 8 times faster per minute than men.
- The number of atoms in an adult human body is equal to 10,000,000,000,000,000,000,000,000,000 units.
- In our body, there are narrow tubes called blood vessels which circulate blood in the body. The total length of human blood vessels is about 100,000 kilometers.
- About 1.5 gallons of blood flowing through the human body circulates through an astonishing network of arteries, capillaries and veins.
- The highest percentage of blood groups in the community belongs to people with O + blood type with 38% percentage.

- The size of white blood cells is about 0.1 mm. There are about 700 red blood cells per each white blood cell.
- Red blood cells contain a pigment called hemoglobin, which is red. So that is why blood looks like red.
- Red blood cells in vertebrates play the role of delivering oxygen to the body. Your body loses 2.5 million red cells every second. Fortunately, at the same time, it makes more.
- Have you ever noticed that when a scratch is made on your body, it stops bleeding after a short time? This is done by blood platelets. Platelets stick to each other at the site of injury to prevent further bleeding. Blood platelets are very small in size and there are about 250,000 thousand platelets in one drop of blood!

- Human blood travels a distance of 19,312 kilometers in just one day!
- The ratio of salt in the blood components of the human body is approximately equal to the ratio of salt in seawater! This reinforces speculation about the beginning of life in the oceans.
- The function of the lungs is to separate oxygen from the air and transfer it to the circulatory system, as well as to separate carbon dioxide from the bloodstream.

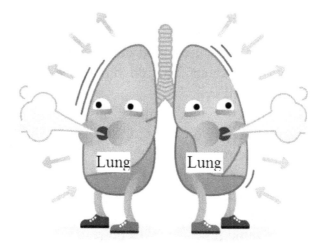

- Humans have two lungs that are located inside the chest. The left lung is slightly smaller than the right lung due to less space due to its proximity to the heart.
- Man stays alive by breathing. An adult human breathes about 23,000 times a day!

- The world record for holding your breath is 7 and a half minutes. Most people can hold their breath for about a minute.
- If water does not reach the body, eventually the lips become wrinkled and black, the tongue becomes swollen so that it does not fit in the mouth and the person almost loses his sight and hearing. The skin turns purple and gray and does not bleed if it is injured anywhere on the body. The kidneys then malfunction and human die after heart failure.

- Up to four kilograms of food accumulate in the human stomach compared to a cow's stomach which can hold ten times this amount of food!
- The human body uses hydrochloric acid to digest food. But the same substance burns the skin. Because the abdomen secretes a kind of mucus to protect itself against acid. When a

man passes away, the production of this mucus stops, the acid begins to destroy the stomach.
- You crave about 500 kg of food a year. This weight is equal to the weight of a small car!
- A large hamburger sandwich has approximately 565 calories. You can burn 100 calories by running for about ten minutes.
- When you vomit, the muscles that normally carry food to the gastrointestinal tract, do the opposite and push food into your mouth!
- The strongest muscle in the human body is the tongue.
- The longest tongue recorded in a human mouth belonged to an Englishman whose tongue was 9.8 cm.

- The human mouth produces about one liter of saliva per day. Saliva helps you crush food before swallowing while keeping your teeth clean.
- Man produces enough saliva in his life to fill two pools of water during his life!
- Everyone's tongue print is as unique as their fingerprint.
- The tongue is one of the organs of the body that shows the health to a large extent. For example, pink indicates good health, white indicates infectious and fungal diseases, and yellow indicates stomach problems or fever.
- Girls have more taste power than boys.
- An 11-year-old child needs about 2,388 calories a day. A bird like a canary needs only 11 calories a day and an elephant needs 91,955 calories!
- People who suffer from migraines (severe headaches) have thicker brains than normal people. Scientists have not yet figured out which one causes the other and what the link is between brain thickness and headaches.
- The capacity of the human brain is ten times greater than previously thought. The storage capacity of the human brain is equivalent to the contents of 4.7 million books or 20 million files full of office documents or 13.3 years of HD quality video or 670 million web pages, or five

times the amount of information in the huge Encyclopedia Britannica.
- The brain becomes lighter over time. The weight of the human brain decreases by one gram per year from the twenties onwards. This is due to the death of cells and the non-replacement of new cells.
- The human brain has one hundred billion nerve cells.
- All kinds of messages in the human body reach the brain at a speed of about 360 kilometers per hour!
- An early human, called Neanderthals, had larger brains than humans today.
- Even thinking takes energy! On average, each human brain consumes as much power as a 10-watt light bulb!

- Your brain receives 100 million units of information from the eyes, nose, ears, skin and other organs at the moment. If scientists wanted

to make a human-like brain from computer chips, they would need a million times as much energy as the human brain!
- Electrical activity is discovered in the human mind, up to 37 hours after death, possibly due to chemical interactions.
- If you pull all the nerve fibers in your brain from the beginning to the end like a wire, you will get a string of 3.2 million kilometers (the distance from Earth to the moon is only 384,000 kilometers!).
- If a human head is beheaded, it will still be awake for a few seconds and even will be able to look around.
- A human's position affects his learning. In other words, human memory is affected by the position of the body, and the person learns things differently while sitting or standing.
- The liver has more than five hundred unique functions in the human body.
- The human fetus has a tail in the mother's womb!
- Even when you are an adult, you still have the same enamel (the most superficial part of a human tooth) that was formed when you were a fetus.
- Babies have more taste buds than adults.
- Most human mutations occur on the Y chromosome which is only found in men.

- Genetic evidence shows that most people in Britain have Spanish ancestry. Spanish sailors are thought to have migrated to and conquered Britain about 6,000 years ago.
- In the eighteenth century, a Russian woman had 69 children. She gave birth to 16 twins, 7 triplets and 4 quadruplets. This is the largest number of children registered in history.

- Newborns do not weep when they cry until they are 6 weeks old.
- The human eye is able to detect candlelight from a distance of 2.76 km.
- The skin of the eyelid is half a millimeter thick (equivalent to a hair).
- Many years ago, a man with four eyes was born in England. His other two eyes were slightly higher than his other eyes. He could close each of his eyes independently of the others.

- The human eye is able to distinguish between 10 million different colors.
- The colored part of the eye is called the iris. The iris muscles contract and expand 100,000 times a day, and if these movements were to be performed on a human foot, we would have to walk 80 kilometers a day!
- Traditional medicine in the field of vision was in the period influenced by Plato's "hypothesis of radius," which said that human eyes emitted rays reaching objects and the return of these rays to the eye caused vision! With this justification, they attributed the refractive errors of the eye (myopia and hyperopia) to something imaginary called the "spirit of vision" and its degree of thinness, concentration or dryness and moisture!
- The human eye is the equivalent of a 576-megapixel camera.
- It takes about 0.2 seconds for the brain to perceive what is being seen.

- The human eye sees the world upside down, but the brain automatically changes and smoothes it for us.
- When you read a novel with 100,000 words (about 350 pages), your eyes move about a kilometer between the pages.

- Genetic research shows that blue-eyed people have common ancestors and about eight percent of the world's population have blue eyes (only two percent have green eyes).
- Eyes can reveal the secret of the heart! The pupil of someone who is interested in you will grow larger in your presence. Although recognizing this requires practice, when you look at someone you love, your pupil enlarges up to 45%!
- 90% of the world's blind live in third world countries, especially Asia and Africa.
- Carrots really enhance vision in the dark. This vegetable contains vitamin A, which helps strengthen the retina.

- Fleas on the body of mice caused a dangerous disease called plague which was common throughout Europe in the fourteenth century,.
- Malaria is a tropical disease transmitted to humans by the Anopheles mosquito. Malaria has been responsible for the deaths of about half of all humans on Earth since the Stone Age!
- Toothpaste is the best ointment to relieve itching in mosquito bites!

- An allergy is a severe reaction of the body's immune system to substances that are usually harmless. Allergies are a common problem among humans and affect about two out of ten people. Cow's milk is more allergenic than most other foods and beverages.
- Laughing seems to reduce the body's allergic reactions. So if you have seasonal allergies, try to laugh at it!

- Kissing is the cause of many disease transmission.
- There is a rare and strange disease called Cotard Syndrome in which a person who is infected thinks that he is dead or has no external existence or, on the contrary, thinks that he is immortal and has eternal life!
- There is a strange disease called pica in which a person has a great appetite for food such as soil, paper and glue and.... Although some believe that this disease is related to mineral deficiency, the cause of this disease or its treatment has not been determined yet.
- "Dancing plague" was a strange disease that occurred in 1518 in Strasbourg, France. A woman started dancing and the others danced with her, and this number reached four hundred. In this strange disease, a person dances to their death!

- There is a kind of mental illness that makes a person believe that a wild animal is like a wolf and behaves like it! This may justify some of the werewolf stories.

- Serotonin is the name of a hormone that causes a pleasant sensation in humans. Foods that contain this hormone include bananas and pineapples, fish and eggs, almonds, yogurt, and milk and cheese. Dopamine hormone also causes a feeling of happiness in humans!
- The hormone adrenaline has been shown to give humans temporary strength!
- Sneezing is a defense reaction in the body to expel secretions during illness.
- Scientists have calculated that the speed of the sneezing process is equivalent to 161 kilometers per hour!
- When you sneeze, your whole body stops working for a moment.
- It is not possible to sneeze with open eyes.

- Cells are the smallest unit of life and are often referred to as the "building blocks of life." The human body produces 1 million new cells every second!
- When you look at your tongue in the early morning, something white covers it. They are cells that have died overnight!
- Dandruff is made up of a combination of dead skin cells and pus and fat on the scalp. You lose millions of cells every day, so there are enough cells for dandruff!
- Up to 50 million bacteria can be present in a drop of water. However, the largest known bacterium is almost visible to the naked eye!
- A number of bacteria are pathogenic in humans. But in general, without the activity of these creatures, life on earth would be disrupted!
- There are more bacteria in your mouth than there are people on Earth! They feed off food debris and dead cells in the mouth.

- The simplest type of reproduction in the world is done in bacteria (double division).
- Bacteria can survive a thousand times more radiation that kills humans.
- Bacteria live inside the pores of active volcanoes, suggesting that life is possible even at high temperatures, and the theory confirms that bacteria are found in all climates and in almost all places, including hot springs, polar ice caps, and lakes with high salinity.
- The human mouth smells bad in the morning, even if it is brushed, because at night, during sleep, the production of saliva stops. As a result, fungal bacteria begin to work and accumulate in the mouth.
- Bacteria in spoiled foods can lead to acute food poisoning. This bacterium is so dangerous that only 450 grams of it can wipe out the entire human race!

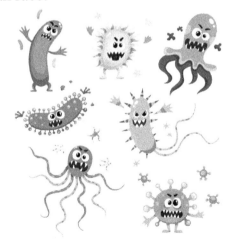

- Do not drink water with "fruit"! Fruits have a moisture that, if consumed with water or after eating, the digestion of the fruit is interrupted.
- Mitochondria are part of the cells in our body. Scientists think they were originally bacteria that were absorbed into our bodies and are now an essential part of us. Their job is to take in nutrients and build energy for our body cells.
- Some foods, like celery, take a lot of energy to chew and digest, which means even more energy than you get from them! So in theory, they are actually reducing your weight!
- Researchers believe that you should not soak your head in water to escape the heat. Pouring water on the head not only does not prevent heatstroke but also will provoke a headache!
- Scientists recommend to avoid placing items in the same place; they say change their location regularly, as this will help prevent Alzheimers!
- The number of toes has decreased due to evolution in many vertebrates, but we humans still have five toes. Since the little toe is useless today, it may disappear in the future!
- If all the molecules in your body are placed in a continuous path, its length is equal to 3,000 trips to the moon!
- Fingernails grow four times faster than toenails.
- People in Pakistan have been going to dentists for 9000 years. Archaeologists have found filled

teeth in the remains of human skulls in this country!
- Eating bread or chewing gum will prevent you from shedding tears when peeling onions.
- The growth of two parts of the human body never stops! The human nose and ears are always growing, even when all the organs of the body have stopped growing.
- One in 20 people has an extra rib.
- One in 10 people is left-handed. But boys are 1.5 times more left-handed than girls.

Terrestrial organisms

- Although the panda is counted as carnivore, 99% of its diet is bamboo!
- Pandas kept at the zoo eat foods such as honey, bird eggs, fish, citrus fruits, bananas, plant leaves and potatoes!
- A large panda can eat 45 kg of bamboo in just one day!
- The panda is now found only in the mountains of China and is endangered.
- The image of a panda bear eating bamboo has become a symbol of peace against predators.
- Bears are known as nature's gardeners! After eating the fruit, they throw the seed miles away from the mother plant, and this causes the plant to grow elsewhere!
- Brown bears may stay in the cave for more than seven months and eat nothing, and then suddenly wake up and eat forty kilograms of food!
- Brown bears almost always avoid encounters with humans.
- A study based on *Wildlife radio telemetry* in Russia found that the main enemy of brown bears are Siberian tigers. During the 20 recorded battles between brown bears and Siberian tigers, brown bears were killed in 50%

of cases, tigers were killed in 27% of cases, and in 23% of battles both animals survived and ended the battle.
- The polar bear, by its size, has the largest food-eating capacity of any animal. This animal can kill a large Walrus and eat it all alone!

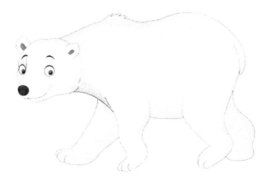

- Polar bears have black skin under their white fur coat!
- Polar bears can smell their prey from about 30 km away, even if they are under ice!
- Infrared camera works with heat and shows every creature except polar bears! Although the inside of their body is quite warm due to the thick layer of fat and thick hair, the temperature of their outer surface is equal to the temperature of the snowy space around them. Hence they are not detected by infrared cameras.
- The Persian fallow deer is a deer species presently merely living in Iran and Israel. This

animal has been listed as endangered on the IUCN Red List in 2008 but after a captive breeding program, the population has rebounded from only a few deer in the 1960s to more than 1,100 individuals today.
- Unlike other horned animals, deer antlers are not permanent and fall off every year and a new set grows in its place!
- In Japanese Shinto, deer are considered messengers of the gods. This animal also has a special place among Native Americans.

- Reindeer are one of the few animals that can eat moss. Moss contains chemicals that help keep the body warm.
- The sloth is a native animal of the America. The name of this animal is given to it due to its slow

movements. In fast mode, they move about four meters per minute.
- A sloth sleeps for about 15 hours a day. Female sloth usually give birth to only one child and their baby hangs on the mother's body for 9 months after birth!

- The hyena feeds on the carcasses of animals hunted by other animals. This animal can chew a broken glass bottle without being hurt.
- Elephants are the only animals with four forward-facing knees.
- The ivory of an African elephant kept in a zoo in New York is 3.49 and 3.35 meters long.
- Although elephants have 40,000 muscles in their bodies, they are the only mammals that cannot jump!

- Among the elephants, the heaviest brain (weighed so far) was 7.5 kg, which belonged to an Asian female elephant of about 3 tons. The brain weight of African male elephants is usually between 4 and 5 kg.
- The lion has earned the title of "King of the Jungle" due to its large physique, agility and awesomeness though they do not usually live in jungles!
- The lion's skull is very similar to that of a tiger, but its front is smoother and has larger airways.
- The main enemy of lions in nature, except humans, is the Nile crocodile. The winner of the confrontation between these two animals is usually fifty-fifty; Lions usually win on land and crocodiles in water.

- When several tigers are on a prey carcass, the male tiger lets the female tigers and their children eat first (unlike lions). They usually wait and take turns eating and are less likely to fight.
- Tigers do not normally attack humans unless threatened. It is said that if you look directly at a tiger, it will not attack you. In some parts of India, ordinary people wear masks on back of their head while walking in the forest to prevent tigers from attacking from behind.

- One of the most famous extinct animal species is the Mazandaran tiger. The last tiger of this species was hunted in Iran in 1952 in an area that is currently located in Golestan province. Despite the extensive research conducted by the Environmental Protection Organization of Iran to find the Mazandaran tiger in the early 1970s, no evidence was found that there was another tiger, and therefore the extinction of the Mazandaran tiger was finally declared.

- The cheetah can reach a speed of 75 km / h in 2 seconds right after starting to move; Their acceleration is higher than many racing cars.
- Cheetahs and leopards both have the same body length. Although many people confuse cheetahs with leopards, the main difference between the two species is in the shape of their spots. In cheetahs, the spots are thick and round, while the leopard spots are large and rosette-shaped.
- The Asiatic cheetah is a highly endangered species of cheetah in that only 12 of them, 9 males, and 3 females, are left in Iran. In 2014, the Iranian national football team announced that their shirts are imprinted with pictures of the Asiatic cheetah in order to bring attention to conservation efforts.
- Cats are one of the cleanest living creatures. They pay a lot of attention to cleanliness. A cat's saliva has a great washing and regenerating power.
- Persian cats along with exotic cats were recognized as the most popular cats in the world in a 2016 poll.
- The Persian cat entered Europe during the 16th century. This type of cat has long hair, a round face and a short nose.
- A cat is more likely to die in a fall from a seven-story building than when it falls from a twenty-story building. This is because the floors give the cat more time to understand the

situation and land with the least possible damage.

- Cats do not walk on aluminum foil, probably because they do not like the sound.
- Cat urine shines in the dark!
- Cats cannot taste sweetness!
- A parasite called toxoplasma prefers to live in a cat's brain, but at the same time it can infect mice and manipulate their brains in a way that makes them less afraid of cats! This complex parasitic plot makes mice more likely to be trapped and eaten by cats, so that the parasite can eventually reach the cat!
- Different types of mice (such as the rat and the house mouse) formed 40% of the mammal population on planet Earth.
- The armadillo is a small mouse-like mammal known for its hard armor. This animal has more

teeth than any other mammal (about 100). However, his food is mostly termites and ants that do not need to be bitten or chewed!
- Researchers have found that shaving pregnant female mice causes them to produce more milk and their babies to grow up. Food digestion and absorption are better in a bald mouse and therefore lead to more milk production.
- The Etruscan shrew (mammals similar to mice but smaller) is so small that it can use a tunnel dug by large earthworms
- There is a type of mouse in the US called the Star-nosed mole. These mice feed on worms, insects and tiny fish. This animal can detect, hunt and eat its prey in a tenth of a second!
- Mice do not get sick. The rat poison that kills mice only works because mice cannot vomit it!
- Using genetic engineering techniques, scientists have created a powerful mouse that can run at a speed of 1.2 kilometers per hour and live longer than a normal mouse. Meanwhile, this mouse has more fertility than normal mice.
- The taste of rat poison varies in different countries! Producers try to produce this toxin that tastes and fits the taste of the mice of that country! It is said that eating 10 grams of black rat poison will cause human death.

- The odor indicator of a dog is 150. This is while the index of a human sense of smell is only 4.
- In 1954, Russian scientist Vladimir Demikhov in fact created a two-headed creature by grafting the front and legs of a puppy to an adult dog! The two heads of the animal sometimes fought with each other. This creature survived for six days. Vladimir Demikhov later created another animal that survived for a month.
- The highest age recorded for a dog was 29 years and 5 months, reported for a Queensland Heeler dog named Bluebird in Australia. This dog served as a *guard dog* for 20 years and finally died on November 14, 1929!
- The fox is very fond of sweet food and even eats grapes and honey and even chocolate, candy and pastilles!

- The giraffe's tongue is so long that it can lick the inside of its own ear!
- The giraffe's heart has a special valve to help pump blood to its full height.
- Early zoologists centuries ago thought that a giraffe was a creature between a camel and a leopard, and called it a camel-leopard!
- Each Zebra has its own unique stripes, and the exact pattern of the lines on the body of no two Zebras is the same. Just like human fingerprints!

- Once upon a time, the population of wolves on earth was several million, but today their population is about 150 thousand. In the past, wolf hunting was a popular pastime and sport among the aristocracy and the rich people. The first paintings of wolves on cave walls in southern Europe date back to more than 20,000 BC.

- According to recent research, about 30,000 years ago, wolves probably reached the edge of human habitat in search of food, and during their battles with them, humans began the process of domesticating and teaching these creatures, which gradually changed their behavior and eventually the dogs of today came into being.

- The strong and sharp rhino horn is made of creatine; the same substance that makes up human hair, skin and nails.
- The skin of rhinos is 3.5 cm thick, they are almost wearing bulletproof clothes!

- Kangaroos can make long jumps with their feet. The length of their jumps is sometimes up to three meters. Kangaroos can only jump when their tails touch the ground.
- Llamas are good guards. Farmers in New Zealand and Australia often use them to guard herds against predators such as foxes.
- Camels have historically been one of the most useful animals for humans. This animal was not only the main tool of human movement and transfer in deserts, but also could help us live for weeks by feeding on their milk and meat and using the hump fat instead of butter and its wool to build tents, blankets, rugs, woolen garments, ropes and cords, shoes, musk. Camel feces were also suitable for lighting fires.

- Koala fingerprints are almost inseparable from human fingerprints and can be mistaken for human fingerprints.
- A hedgehog can swallow 100 times more toxic hydrogen cyanide, which is deadly to us, than a human, without harm.
- Hedgehogs can stay afloat on water!
- Malagasy hedgehogs are very greedy. They often eat enough to get sick!
- The platypus is a mammal that lives mostly on the Oceania continent. This animals have beaks similar to a duck's beak, which has sensitive nerves and help them find food. A sharpener can detect electric current with its beak!

- Some scientists believe that the first animals gradually came to land from the sea and grew feet once there.
- The size of a frog in America is bigger in childhood than in adulthood!
- Using high magnetic force, it is possible to lift a small frog off the ground and hold it in the air.
- Frog's legs is one of the most popular foods in many parts of the world. This food is consumed in countries such as France, China, Thailand, Indonesia, Vietnam, Portugal, Spain, Albania, Slovenia, Greece, India, Italy and the United States, etc. Frog legs are rich in protein, omega-3 fatty acids, vitamin A and potassium. According to Islamic dietary rules, eating frog meat is forbidden.
- Japanese scientists have developed glasses that can be used to see the internal organs of a frog. So there is no need to slaughter them for autopsy.
- When a frog gets sick, it brings out its entire abdomen, which hangs from its mouth. A frog

uses its arms to expel the contents of the stomach and then swallows the stomach again!
- The tree frog is pale green in sunlight, but changes color and turns white when it enters shade! This frog often lives among the trees and in the toilets! Yes, do not be surprised, you can find this type of frog in the toilets of Australia and New Guinea!
- Among amphibians, the largest animal is the Chinese giant salamander. This amphibian lives in the cold mountain rivers of northeastern, central and southern China and the surrounding swamps. The average length of an adult amphibian is more than one meter and it weighs between 20 and 30 kg.
- Do not be surprised if you encounter an open-mouthed crocodile during your visit to the zoo. Their open mouth is not a sign of their anger. The crocodile does not have sweat glands, so it keeps its mouth open in summer to cool the temperature inside its body!
- The muscles that help the crocodile keep its mouth open are very weak, so humans can hold it closed with their bare hands! However, it is better not to try this!

- The crocodile cannot stick its tongue out of its mouth.
- Mata Mata is a South American native tortoise that usually lives near freshwater. This animal walks so slowly that algae grow on its lacquer!
- The Galapagos Giant Tortoise was a species of giant tortoise that is now extinct. This group of turtles had a long life. An Australian tortoise named Harriet, who died naturally in 2006, was 170 years old. Harriet is said to have been discovered in 1835 by Charles Darwin.
- The oldest turtle fossils were found 200 million years ago (Jurassic period). Scientists believe that turtles have not changed much throughout history due to their physique.
- The largest snake in the world is the South American Green Anaconda. This animal is a member of the boa species. Green anacondas can grow up to about nine meters in length and weigh up to about 250 kilograms.
- It is rare for a snake to be born with two heads! In that case, they are fighting over food!

- People generally confuse the legless lizard with the snake. Unlike lizards, snakes do not have eyelids or holes for the outer ear, their scales are broad, and they have claw tongues. They also often have long limbs, unlike lizards, and have short tails.
- The rattlesnake venom remains fatal for 25 years after the animal dies!
- The longest snake bite belongs to the Gaboon viper, a snake native to tropical Africa. The size of this snake bite is about four centimeters.

- The heaviest venomous snake is the diamondback rattlesnake found in the southeastern United States. This snake weighs 15 and a half kilograms.
- Eurasian vipers can live for a year without biting anything!
- On February 12, 1962, in the Philadelphia Botanical Garden, a viper strangely committed suicide by stabbing itself in the back!
- Laurence Monroe Klauber, scientist and biologist (1883-1968), is known as the Rrattlesnake! He collected 35,000 snakes and reptiles during his lifetime.
- An Asian lizard that had been frozen for 90 years came to life after the ice melted and set off! The strange thing is that this animal lives only 10 years and goes into hibernation during the winter.

- Lizards can move their eyes independently, so they can look in two directions at the same time.
- Lizards are able to regenerate lost or damaged parts of their body, including their legs, eyes, and even their heart! Scientists believe that perhaps genetic engineering can do the same thing in the human body.
- The size of the eye is kind of related to the speed of that creature more than any other organ! Faster animals, such as brown rabbits, have larger eyes than slower animals. On the other hand, animals that move slowly, such as blind mice, have very small eyes or in some cases no eyes at all, and of course, the life of animals underground is another factor that affects their speed.
- There are about 10 million species of living creatures on planet Earth. We humans are just one of them!

- It is estimated that only 10% of the species today have existed since the beginning of history and the rest are extinct.
- The only animals that can detect their reflection in the mirror are monkeys, dolphins and elephants.
- Babirusa or Deer pig is an Indonesian pig with curved branches. These branches grow from its nose. As the animal ages, it sometimes spins its branches and gets stuck in its jaws. Undoubtedly, the design of this animal was not good!

- Carnivores do not eat an animal that has died from lightning or thunder!
- An anteater can eat 30,000 ants a day!
- Coconut crab is the largest crab in the world, sometimes reaching one meter in size! This crab is famous for eating coconut. The power of coconut crabs in biting is greater than any other animal on earth, including leopards, lions, bears, wolves, etc!.
- The crab is an animal of the crustacean subfamily. Crab skin is made of chitin, the same substance that mushrooms are made of!
- Crabs are omnivorous; their diet includes algae and other foods like mollusks, worms, other crustaceans, fungi and bacteria, and any small food available to them.
- Before us, the original inhabitants of the planet were creatures called dinosaurs. These giant reptiles became extinct about 65 million years ago. So far, different species of them have been discovered through their fossils, but researchers believe that 70% of the dinosaurs have not yet been identified.
- One of the giant dinosaurs called Sauroposeidon could stretch its neck up to 17 meters!

- One herbivorous dinosaur called Stegosaurus was 9 meters long and 4 meters tall, yet its brain was only the size of a walnut!

- Scientists have found the remains of a 1.8-meter dinosaur that tunneled underground like a giant mouse.
- A giant herbivorous dinosaur called Diplodocus had a tail 14 meters long.

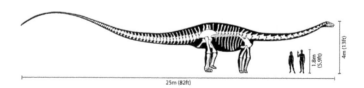

- Tyrannosaurus Rex, or the tyrant lizard king known as T-Rex, looked like a turkey with a fur at birth, despite its horrible appearance. However, it is believed over time they lost their feathers. Within two decades, T-Rex transformed from a chicken into a 9-ton carnivorous monster.

Emperor of the Sea

- The blue whale has the slowest heart rate of all animals. The animal's heart beats 4 to 8 times per minute.
- Scientists can detect the sounds produced by whales from a distance of 850 km with the help of sound wave detection tools!

- A baby blue whale drinks 227 liters of milk daily and grows 91 kg every day!
- The whale has the largest egg among all living animals. On June 29, 1953, a fisherman named Doris managed to catch an egg measuring 30*14*8 cm from a depth of 57 meters in the Gulf of Mexico. Inside the egg was a full 30 cm long whale embryo.

- Whales cannot move their pupils. They have to move their whole body to change the direction of their gaze!
- A 24-meter whale called the cachalot has the largest brain in the world, weighing up 7 to 9 kilograms.

- A shark is like a cat designed to hunt. They detect vibration at a distance of about One kilometer and detect a drop of blood in a volume equal to 25 gallons of water!
- Sharks must swim to get water out of their gills. If they stop, they will drown!
- In 2001, a stray shark gave birth alone (without a mate) at a zoo in Nebraska, USA. Some bony fish can also have children on their own.
- An estimated 100 million sharks are hunted by humans each year, reducing the population of stray sharks in the ocean by 98 percent.

- Although Steven Spielberg's 1975 film *Jaws* depicts a white shark as a wild human-eater, the predator's main food is marine mammals.

- Electrophorus electricus is an aquatic creature that can produce strong electric shocks up to 860 volts to hunt and defend itself! So almost no hunter dares to step forward to try eating this fish. At a minimum, this amount of electricity can numb any animal or deter it from attacking. These fish are more than 2 meters long and weigh up to 20 kg. Another weird thing is that these eels can also control the power output! For example, use lower currents to hunt and stronger currents to defend themselves.
- With the amount of electricity produced by the electric eels, two electric refrigerators can be kept on at the same time!
- Cusk-eel live more than eight kilometers below sea level!
- There is an eel called Gapler that goes down to 5 km in deep water. The mouth of this strange

animal opens up to 180 degrees, which allows it to swallow prey even bigger than itself!
- Trouts use a geographic magnetic map to find their way from the ocean to their native river.
- Dolphins are able to exchange voice messages. Some of their hearing sensors can receive sounds up to 200 kHz. In this way, they can send and receive vital information to each other over long distances.
- Dolphins are very interested in humans for unknown reasons. They help fishermen catch more fish, help frogmen find sunken ships, rescue people who are drowning, and cooperate with humans in aquatic activities.
- The goldfish or red fish, which as traditionally part of Nowruz and is used every year on the table of Iranians (called Haft-sin), is the most famous aquarium fish in the world and more than 480 million of them are sold yearly.
- It is more like a myth to say that goldfish have only a few seconds of memory. To study this, scientists have taught goldfish that food is prepared to eat with the sound of bells. Scientists then released them into the sea. After five months, playing the same bell brought the goldfishes to the surface!
- Goldfish are the only animals that can see in both ultraviolet and infrared light. If you leave these fish in a completely dark room, they will eventually turn white.

- Many people in Iran think that goldfish have a short lifespan (a few days)! But it is interesting to know that the primary cause of their death is the poor care of humans. The record of the oldest goldfish was more than forty years!

- Some species of deep-sea creatures emit a glowing liquid. This creates a cloud of dazzling light and distracts the hunters!
- The walrus can be sunburned! Other creatures that get sunburned other than humans include white horses and pigs.
- Some scientists believe that "silent" genes can come back after millions of years. With the return of these genes, we may see dolphins with legs, or bears with common aquatic fins.
- Coelacanth is a rare species of fish that is considered to be the oldest living carnivorous fish in the world. Scientists thought the species became extinct about 65 million years ago until a healthy specimen was found in 1938.

- The Lion's mane is the largest type of mermaid in the world and is very dangerous. The body diameter of this aquatic animal is 2 meters and the length of its legs is 15 meters.
- One of Sherlock Holmes stories called "The Adventure of the Lion's Mane" is about the same animal. If you put the tentacles or so-called manes of this animal end to end, they will cover an area the size of fifteen tennis courts!

- The scallop has 100 eyes near its shell. These eyes help to identify the shadow of the hunter approaching it.
- Some fish that live deep in the sea and in places that are always dark use chemical reactions to produce light by using the phosphorescence process.
- The European flounder is a type of deep-watrer fish. In childhood, this fish, like normal fish,

has eyes on either side of its head, but as it ages, one eye moves to the other side, so in adulthood, both eyes are on one side of its head. So these fishes keep only one side facing up and the other side always towards the sea floor.
- Temporary lagoons in Venezuela are home to fish that can survive in sludge for more than 60 days without water and oxygen once the water has dried up. No other animal can hold its breath so long!
- A sailfish is a type of fish that has a stretched and compressed body and its dorsal fin is large and looks like a sail. This fish can swim faster than a cheetah. The swimming speed of this species of fish is at least 109 kilometers per hour, while the maximum speed of cheetahs is about 100 kilometers per hour.
- Fish can also become seasick.
- A tuna never stops. These fish move throughout all their lives, traveling about 1.6 million kilometers in 15 years.
- A huge squid fish caught in Canada in 1878 had a body six meters long and tentacles 10 meters long!
- The catfish is a type of fish that is found in different parts of the world, including the Persian Gulf. The reason for naming this fish is the whiskers around its mouth. The taste ability of this creature is ten times stronger than humans!

- The viperfish is a carnivorous fish. This fish attacks its prey directly. The needle-like teeth of this fish are so large that they do not fit completely in its mouth!
- Another name for caviar is sturgeon.
- 93% of the world's caviar fish reserves are in the Caspian Sea.
- Carp go into hibernation en masse at temperatures below 7 ° C!
- The main source of food for deep-sea animals are the bodies and feces of other creatures that live at higher water levels!
- Xiphophorus is one of the most popular aquarium fish. The fin of this fish extends in the shape of a sword. Each of these fish has its own unique personality and temperament! That is, one of them may be a cowardly, shy fish or an aggressive, angry one!
- The starfish is an aquatic animal that can only live in brackish waters and therefore does not exist anywhere like the Caspian Sea. The weird thing about a starfish is that it has no brains!
- Without starfish, the oyster population (which is the primary food source of starfish!) Will increase to the point of explosion and the situation will be such that the balance of all waters and oceans will be disturbed.
- There is a kind of oyster in the depths of the North Atlantic that grows only 8 mm in a hundred years.

- Papua New Guinea is a Pacific country in northern Australia and eastern Indonesia. Until 1933, the currency of Papua New Guinea were oysters!
- An oyster recently found off the coast of Iceland was between 405 and 410 years old. This animal is the oldest living thing on earth. Researchers named the oyster Ming. When Abbas the Great ruled in Iran, the creature was considered a kid!
- Oysters can change sex several times in their lifetime!
- Some octopuses pick up the bites of the starfish they have hunted and use it as a weapon!
- An octopus is an animal that has no bones in its body. When the octopus becomes stressful, it starts eating its own legs!
- The octopus has 3 hearts and its blood is blue!

- Some experiments show that the octopus has extraordinary intelligence. This creature has both long-term and short-term memory and is able to solve puzzles and recognize different shapes and patterns! They use very innovative methods to escape from predators.
- If one of the octopus arms is amputated for any reason, the other arm grows instead!

The Shah of heaven

- It is never difficult for crows to find food. They can eat everything from plants and fruits to the meat and carcasses of dead animals and even garbage.
- Sometimes a crow knocks itself to a window so hard that it hurts itself! Some believe that because the crow sees the reflection of its own image, it thinks it has encountered a rival and tries to fight it.
- Crows can untie knots and undo zippers!
- Research has shown that a crow attempting to eat fruits such as hazelnuts or walnuts (which are not easy to open), places them in the middle of the street for a car to pass over and then eats the contents!
- The anatomy of a crow's brain is somewhat similar to a human brain, and the crow's language is very complex. Every croaking sound of this animal has a different meaning. Some of their sounds are alarming, some are imitations of other living things, and some are related to a specific event. A 2004 study found that crows were smarter even than monkeys.
- When you walk past a crow, do you think you can recognize it a few minutes later? Is it really possible to easily distinguish one crow from

several crows? Most likely not. But the opposite is true. Each of the crows you pass by will no doubt remember your face! Researchers at the University of Seattle put on scary masks during their research and threw small stone at 7 crows. They then released the crows. A few days later, each time a member of the group appeared with the same mask, not only those 7 crows but also the rest of the crows in the neighborhood attacked the masked person! The experiment not only revealed that crows would remember faces, but also that they were vindictive.
- The eagle is so scary that other birds, as soon as they see it, run away to save their lives and get away from its flight area!

- Bald eagles can fly 3,000 meters above the ground.
- Some eagles can fly at a speed of 160 kilometers per hour!
- The duck is one of the first species of birds to be domesticated by humans, and it seems that this bird was first domesticated in Egypt.
- The older the ducks, the whiter their color becomes. Of course, white ducks stay white throughout their lives, but dark ducks gradually grow white feathers as they get older, and eventually, when they get very old, their feathers may turn completely white like human hair.
- Sea ducks become seasick if kept in a boat!
- A duckling thinks that any creature it sees when it first opens its eyes will always be the mother.
- Geese are intelligent and loyal birds and can be trusted to guard the house like dogs. These birds start making noise when they sense anything suspicious.
- The most popular goose among food manufacturers is the Toulouse goose. This type of goose is of French origin and is notable for its high volume of meat.
- The homing pigeon uses the Earth's magnetic field to find the path of its nest.
- From 3,000 BC, the Egyptian pharaohs used pigeons as messengers and even considered them sacred birds. The Greeks, Persians,

Chinese, and Romans also used pigeons as messengers for thousands of years.
- A pigeon's bones weigh less than its feathers and wings.
- In 1810, the passenger pigeon was considered the most populous creature in the world. The irony is that this bird became extinct in 1914, about a hundred years later.
- Cher Ami was a domestic pigeon used by the US military during World War I. This bird carried Allied letters despite numerous injuries. Cher Ami officially received the Cross Medal after the war!
- Male peacocks have eye-like designs on their feathers when they open their wings, which is why they are known as thousand-eyed birds in Persia.
- Peacocks have no pigments and their feathers are actually transparent, but the reflection of light produces colors.
- Peacocks are omnivorous, meaning they feed on plants and vegetables, grass, grass, insects, small snakes, and ground-dwelling lizards!
- Peacocks are considered to be one of the biggest birds in the world thanks to their feathers that grow up to 2 meters!
- The peacock makes a scratching sound despite its pretty appearance!

- A type of bat in West Africa lives in spider webs.
- None of the bat species are scientifically classified as birds. Bats are mammals.
- Bats are the only mammals that can fly!
- The flight of bats is different from the flight of birds in many ways. They float in the air, which makes them more maneuverable than birds, capable of catching prey in flight and changing their speed and direction.
- Horseshoe bats in India allow hedgehogs to use their habitat as well!
- Bats always go left when they come out of the cave!
- There are more than 1,400 species of bats in the world, only 3 of which feed on blood. The territory of this group of bats is mostly in South America.

- Bats can fly at high speeds and do not encounter any obstacles in their path. They are able to recognize the reflected sounds of objects around them and interpret it as an "audio image"!
- Owls are unable to rotate their eyes, so they have to turn their heads to see around. Their eyeballs are not round and spherical, unlike other creatures, but they are long and tubular and are fixed in place by strong bones.
- Owls sometimes even turn their heads up to 270 degrees to monitor the environment. Such a rotation of the neck in any other living thing interrupts blood circulation.
- Owls swallow their prey completely and vomit bones, hair and other indigestible parts after digesting food.

- If a chicken gets caught in a tornado, it will lose its feathers, though it will eventually survive!
- Chickens and roosters have more than 30 different sounds and each sound has a special meaning.
- Chickens and roosters can taste salty flavor but not sweet.
- A rooster that was beheaded incorrectly in 1945 survived for 18 months without a head! The owner of this rooster, who was a farmer, called this helpless rooster, Mike, and set up a tour all over the country, exposing the rooster to public display and making money this way.
- A bird called the bar-tailed godwit can fly non-stop from Alaska (near the North Pole) to New Zealand. The distance of this route is about 11500 km. This bird loses 55% of its body weight along the way.

- Instinct in birds is so strong that birds in cages at a certain times unconsciously like wild birds try to fly in the normal migration direction! Even artificial light and heat cannot deceive them.

- Hummingbirds cannot walk. They can only land and fly.
- Among all birds, the hummingbirds has the lowest number of feathers with about 1000 feathers. The swan has the largest number with about 25,000 feathers.
- A pelican can hold up to 2.5 gallons of water in its bag-like beak at any moment.
- The only bird with a nostril in its beak is the kiwi. This bird can smell its food on the ground through holes.
- Woodpeckers, unlike other birds that make sounds, can identify each other by the sound of their beaks pounding on wood!
- A green woodpecker can eat up to 2,000 ants a day.

- Falcons can see a 10 cm object from a distance of 1.5 km. The bird's visual acuity does not change even when flying in the sky at a speed of 160 kilometers per hour.

- Parrots do not talk through vocal cords. They have no vocal cords at all and only regulate the airflow in specific ways by setting their throat muscles to mimic different sounds.
- The most talkative parrot in the world is the African Gray Parrot. This bird can learn more than 80 words, while Casco species can learn up to 50 words.
- The Kākāpō known as the owl parrot is the biggest parrot in the world. The lifespan of this bird is about 95 years. However, it is endangered and by 2018, only 149 of them have been sighted.
- The results of a study showed that parrots have the same logical powers as a 4-year-old human child.

- In a strange move, Conure parrots start plucking their feathers one by one when they are not noticed. This action means they require attention.
- You must have heard that some birds lay their eggs in other birds' nests. They frequently visit the nests and guard their eggs. In some cases, it has been observed that when the host drives out their eggs, the bird attacks host eggs and even kills its chicks!
- About three-quarters of the Earth's birds age less than a year. Surprisingly, in most cases, the larger the bird, the longer it lives (some birds live up to 80 years).
- The dodo was a bird on the island of Mauritius in the Indian Ocean that could not fly. This animal became extinct in 1681. The dodo was 1 meter tall and weighed twenty kilograms. The dodo was not afraid of humans. This feature, in addition to its inability to fly, made it an easy target. Humans also took with them domestic animals such as dogs, cats, and pigs, which also played a role in the bird's extinction.
- Today, the dodo is recognized as symbol of animals that became extinct due to direct human activity, and many environmental organizations use the dodo image and logo to promote the conservation of endangered species.

- Cassowaries, which are known as an attacking bird, can kill an enemy with one strike. Some experts considered them as one of the last surviving dinosaurs.
- There is a strange bird in South America called the hoatzin, also known as the reptile bird. This bird has a fork on its wings and uses it to climb trees. The strange thing about Hoatzin is that its digestive system is similar to that of cows.
- The emperor penguin is the biggest species of penguin in the world. This animal lives in Antarctica and can dive to a depth of 534 meters underwater; That is, 1.5 times deeper than the depth recorded by humans underwater diving.
- Although penguins cannot fly, they can jump up to two meters in the air!
- Australian ostriches that lived in Australia between 15 million to 24,000 years ago were 3 meters tall and weighed 500 kilograms! Ostriches have very small brains compared to their body size.

- The toucan is a funny bird with a very large beak. Its beak is sometimes about two-thirds the length of its entire body! The bird's tongue is smaller than its beak and therefore cannot be used to help deliver food to its mouth.

- Birds' lungs are more complex than mammals' and occupy a large volume of their body. For example, one-twentieth of the human body is made up of lungs, but this amount is one-fifth for birds.
- Some seabirds have some red oil in their eyes, which acts as natural sunglasses and protects their eyes from the sun.
- The black tern can fly in the air for 10 years and drink and sleep in the flight mode! These swallows only land on the ground to raise their chicks.

- The tailorbird uses its sharp beak as a needle, and instead of thread uses vegetable or grass to make the nest.
- The vulture is a carnivorous bird and feeds on the carcasses of other animals and even humans. An Egyptian vulture throws stones at ostrich eggs to break them and then feed on the perforated eggs.

The small creatures

- When a female dragonfly becomes pregnant, it pierces a reed stalk in water with a thin tweezers at the end of its tail, laying a few very small eggs in the middle of it. After a few weeks, the baby dragonflies hatch.
- The baby dragonfly has special tools for swimming. No water creature has such a tool. Humans have succeeded in building jets, but at least no jet submarines have been built so far. The baby dragonfly swims like a jet submarine.
- Dragonflies have a 360-degree field of view! A dragonfly can also fly at speeds of up to 58 kilometers per hour!
- The Australian dragonfly, called Odanta, is the fastest flying insect.
- Bees communicate with each other by dancing and wagging their tails. Ironically, however, bees have different languages in different areas and there is no common standard.

- The bee flutters its wings about 230 times per second while flying!
- In terms of chemical composition, bee stings belong to the snake venom family, and 20 bee stings will lead to the death of a human in a short period of time. Horses are even more sensitive than we are. Six bee stings are enough to kill one. However, beekeepers gradually become immune to bee stings.
- The bee breed has nothing to do with the taste of honey. The taste of honey depends on the flowers that the bee has collected the nectar of from them.
- Worker bees have 5,500 lenses in each eye.
- The light produced by six fireflies is enough for a human to be able to read a book!
- About 4,400 species of aphids have been identified so far, of which only 250 species

have been categorized as serious pests for agriculture and forestry.
- An aphid gives birth to several offspring and each offspring has several developing embryos in its womb! Aphids do not need males to reproduce, and during one summer one aphid can give birth to countless baby aphids!
- American cockroaches are among the insects that many people hate. You may have wondered many times what the reason for the creation of these creatures *was*. It is interesting to note that the appearance of these insects on Earth dates back to at least 300 million years ago. This means that they stepped on the earth long before us!
- The beetle can live for more than a week after being beheaded, and it is ultimately starvation that kills it!
- If a American cockroach loses one leg, another leg will grow in its place.
- The number of squeaks that crickets make is mysteriously based on air temperature! They are so accurate that air temperature can be calculated based on their squeaks.
- Crickets hear through their legs.
- The cricket sound is made by rubbing the wings of this insect together.
- In Cambodia and Vietnam, eating crickets is common!

- Copepods are found in almost all freshwater habitats. They live in groups and sometimes their number in each group reaches 100 trillion.
- The most dangerous animal for humans is the housefly. This insect transmits diseases more than any other creature.
- The foot of a housefly has ten million more times the sense of taste than human tongue!
- 200 million flies can move a car at a speed of 64 kilometers per hour!

- The fly's eye receives and processes landscapes five times faster than human. So when we play a movie for a fly, the number of frames needs to be 5 times faster, otherwise the movie will look like a collection of slow images for the fly!
- The compound eye of a fly is made up of thousands of tiny eyes, and the number of tiny eyes reaches 4,000, which are uniformly square. These eyes allow the insect to look in all direction at the same time.
- Another strange thing about flies is that they are one of the most widely dispersed insects, and there an incredible number of more than 750,000 species been discovered.
- Termites live longer than other insects. The termite queen lives up to fifty years and during this time she lays an incredible number of one egg per second!
- Ants recognize members of their own community by a sense of smell!
- Scientists have recently discovered that climate change, such as global warming and melting ice, affects ants' social behavior!
- The weaver ant sews the leaves together, it does not use a thorn or a needle to do this, but uses its babies. it forces the baby to make a thread out of his mouth, pulls it back and forth, and passes it through the leaf holes like a needle!
- The ant stretches when it wakes up in the morning!

- It is estimated that more than 1 million ants live on Earth per each human being. The weight of all the ants on Earth is almost equal to the weight of the entire human population!
- The head of an ant is larger than its own body, and this ratio is higher than any other animal.
- Ants are incredibly powerful. From the point of view of scientists, the strength of an animal should be measured by the scale of its weight and size. By that logic, the strongest animal on earth is an ant called the Asian Weaver Ant, which can lift objects up to 100 times heavier than its own weight and move them easily.

- About 13,000 species of ants have been discovered around the globe so far. The length of these ants varies from 1 to 30 mm and their maximum weight can reach 30 mg.
- A type of trumpet tree in South America has protective ants on its trunk! These ants, which are in the category of Aztec ants, bite any

foreign organism that is on the tree surface and then secrete an acid that aggravates the pain.
- The biggest spider ever discovered is the Goliath birdeater, found in 1965 in Venezuela. This spider was 28 cm!
- The black widow spider is common in the south of England. At first, this spider did not exist in that area at all, but it came to England with shipments of bananas, and the suitable climate of the region caused this creature to live there!
- Some types of Australian baby spiders bite their mother's feet and eat them within a few weeks!
- Tarantulas are a type of spider, some can measure up to 1 cm in size. Most tarantulas live in desert areas and feed on arthropods and small vertebrates.
- The largest spider ever found was Megarachne, the fossil of which was found in Argentina. This spider was 50 cm long!
- There are 35,000 species of spiders, but only 27 of them are capable of killing humans.
- Most people know spiders because of their webs, but some spiders do not produce any web. For example, there are spiders called wolf spiders that lurk without web and hunt their prey, and there are jumping spiders that do not need a web due to their strong eyesight and high speed. They simply attack their prey.
- Spider webs have high strength and are also extremely lightweight and flexible. If we make

a net from spider's web, with that net we can stop a Boeing 747. But it takes 27,000 spiders to make about half a kilogram of threads!

- Insects eat 10 percent of the planet's food resources annually.
- An average of 10,000 new insect species are discovered each year!
- Earthworms are soft animals and have no bones in their bodies. An earthworm eats as much food as it weighs daily and produces fertilizer. There is a kind of sense system in the body of the earthworm that it can taste things even with his hands and feet!
- The total weight of all earthworms living in the United States is fifty times the weight of all humans on Earth!
- If a healthy earthworm is halved in the right place, it can make a new head or tail!

- The ribbon worm has a length of about 20 cm. This animal can eat 95% of its body without dying! Of course, it does this only when he cannot find anything to eat!
- Planaria worms can even grow from a small part of themselves! That is, if we divide it into 5 parts, after a while, it turns into five new Planaria worms!
- It is interesting to know that worms do not chew their food but swallow it whole. Small stones are then swallowed to crush the leaves and other plant material they have eaten. There are other animals that eat stones to digest food, even dinosaurs did.
- You may not believe it, but 80% of all living creatures on earth are worms! These worms, known as roundworms, are ubiquitous, and can be found from salt and fresh water to soil and plants to the bodies of other creatures!

- Leeches are a kind of annelid. This strange animal has 32 brains! The body surface of the leech worm has about 102 rings. Another strange thing about this worm is that it has three jaws, each of which has 100 teeth. In other words, a leech has 300 very small teeth!
- A leech does not need to eat most of the time! Sometimes the cells that it ate from a meal 18 months ago are found in its stomach and the same food satisfies it!
- If the discussion is about overeating, leeches will overtake many animals! This creature is able to suck blood five times more than its own body volume in just twenty minutes!

- Leeches live in swamps, paddy fields and moors. Their activity is usually in the dark areas and they hide in the light. So hunters when it gets dark, wear special clothes and go to those areas where leeches are found. In such a situation, the leeches come out and cling to the hunter's clothes to suck blood, not to mention

that they have been hunted by the hunters themselves!
- Some people attract mosquitoes more because of chemicals in their skin. According to researches, 277 out of the 346 fragrances in human hands (created from various chemical compounds) will attract mosquitoes.
- One of the substances that wards off mosquitoes is eucalyptus oil.
- The largest group of grasshoppers ever seen was in Nebraska, USA in 1875, when more than 1,000 billion winged grasshoppers flew in groups!

- According to some Islamic Marja, eating locusts is halal.
- On the way Iranian city, Yasuj, there is a waterfall called Ab Malakh (water grasshopper). According to the locals, the water

of this waterfall has compounds that kill pests, especially locusts.
- In 2000, more than 100 billion locusts attacked barley and wheat crops on Australian farms.
- Locusts can jump 500 times their own height. If man could jump like a locust, he could easily jump as far as a football stadium or even farther!
- In 1915, large numbers of locusts darkened the skies of al-Quds (Jerusalem) in the middle of the day, blocking the sun's rays! Locusts ate everything and laid millions of eggs. Although all 15 to 60 year-olds people were forced to collect and destroy 5 kilograms of eggs, 99 percent of the locusts hatched.
- Butterflies have taste receptors on their legs that help them find the right leaves on which to lay their eggs!
- A butterfly has about 12,000 eyes.
- Butterflies can produce high frequencies that prevent them from being found by predators such as bats.
- The viscous substance of the snail's body is so sticky that it can crawl on the edge of an extremely sharp razor and not be harmed at all.
- The number of insects on the planet Earth is 100 times more than the number of humans!

- A scorpion can tolerate 200 times as much radiation as humans can suffer. Scorpions are highly resistant to hunger and thirst and can survive for months without water or food.
- Of about 200 different species of scorpions around the world, only about 20 are poisonous.
- Some 390 million years ago there were sea scorpions that were larger than humans and lived in swamps.

Mother Earth

- The deepest lake in the world, Lake Baikal, is located in central Siberia. The length of this lake is 619 km. In 1957, the depth of the lake was reported to be 1940 meters!
- The Caspian Sea with about 400 thousand square kilometers is the largest lake in the world. The deepest point of the Caspian Sea is near the coast of Iran and has a depth of about one kilometer.
- Different ethnic groups call the Caspian Sea by different names. The Greeks called it the Hyrcanian Ocean. The Indians called it Kashiab Sagar. In Turkey (and in recent decades in Iran) it has been referred to as the Khazar Sea. Its official and international name is Caspian, which refers to an ethnic group called the Caspi. According to Russian historians in the twelfth century, this sea was called Tabarestan (AKA Mazandaran) by the Iranians.
- Between 40 and 44% of the water of lakes around the world is in the Caspian Sea. However, the water in its northern regions is very shallow near Russia, and only half a percent of the water is in the northern quarter (although it is worth noting that due to the depth and slope of the waters towards Iran, southern beaches are more polluted than anywhere else).

- The Caspian Sea has many scattered islands, most of which are uninhabited. The largest island in the Caspian Sea is the island of Ogurja Ada and belongs to Turkmenistan.
- Diamonds are actually crystalline grains of carbon and are considered precious stones. Crystalline grains of carbon are formed only when the carbon is hard pressed and heated at the same time. By that logic, it is possible for diamonds to fall on other planets instead of rain, because the atmospheres of planets such as

Neptune, Uranus, Jupiter, and Saturn have such high pressures that they are likely to crystallize carbon atoms into diamonds!
- Diamonds are not as rare as they seem, but due to high demand, they have kept their prices artificially high.
- Large cacti found in the Arizona desert grow less than 2.5 cm in the first 10 years of life.
- The Saguaro cactus lives up to 200 years and grows up to 185 meters. This cactus can store up to 8 tons of water, but if you are left alone in the desert one day, do not cut its stem to drink water, because it is poisonous to humans!

- Lightning is a type of electrical discharge between two clouds or clouds and the Earth. The number of lightning strikes around the world is between 1500 and 2000 times per second!

- The temperature in the place of lightning reaches about 27,500 degrees Celsius, which is about 5 times the temperature of the sun's surface!
- Lightning does not always come down from the sky to the ground, but in some rare cases it does the opposite! In this case, the Earth has a negative charge and the cloud has a positive charge.
- In the past, it was believed that lightning never hit a point twice, but today, science says that there is no reason lightning should not hit a place twice, and for example, on average, more than 100 lightning strikes hit the Empire State Building in New York, USA, during the year.
- Skyscrapers around the world use devices called lightning rods which prevent lightning damage to the building and even store electricity to be used in the building's electricity!

- Lightning kills about 2,000 people yearly!

- Ball lightning is a phenomenon that is sometimes seen as mass of intense light in the form of a sphere for a few seconds during thunderstorms. No one knows exactly what causes this phenomenon, and despite numerous observations by scientists, there is even doubt about its existence. In 1994, Ball lightning left a hole in a closed window 5 cm long.
- Lightning is known by various names, including thunderstroke, levin, thunderbolt, electrical discharge, firebolt, thunderball, bolt-from-the-blue and streak of lightning.
- In the past, nectar from bellflowers was used to make glue.
- In the past, the rose was introduced as the god of love and affection. With the advent of Christianity, this flower became a symbol of piety. Gradually, however, the rose lost its symbolic value and was used more in funerals!
- The rose is the most popular flower in the world, and it is also a great source of vitamin C!
- If you put any flower in water containing ink, it is likely that the flower will turn dark blue!
- The worst flower in the world is called the Titan arum, which smells like a decaying animal. This flower is found in Southeast Asia and is sometimes up to 3 meters high.
- For several centuries, Norwegian angelica was used in Europe to cure diseases (from plague to

indigestion). People believed that this flower repels evil spirits.
- We have about 1,500 species of irises. The iris plates that grow on the waters of the Amazon are so strong that a child can sit on them!
- A type of iris, called Chelcheragh iris, grows only in the provinces of Ardabil and Gilan in Iran and the Lankaran region of the Republic of Azerbaijan (and nowhere else in the world).

- A type of Cycad that was purchased in 1775 by the English Kew gardens (the famous flower and plant garden in England) is still in its place after more than 240 years!
- There is a rare plant in Bolivia that you have to wait between 80 and 150 years before its flowers bloom!
- The 2,000-year-old seeds of the water lily will grow if planted in a suitable place!

- Nettle is a common plant that grows in humid areas. This plant grows best in soils where corpses are buried. This is because of the phosphorus chemicals in the bones that help the plant grow.
- Orchid seeds are so light that more than 1 million units weigh only about 1 gram.
- The Venus flytrap is a carnivorous plant that traps and eats insects. However, its operation is slow and it takes half an hour to suffocate and kill a fly, and it takes 10 days to digest it. Venus flytraps often grow in areas with low sources of nitrogen, so it tries to compensate for the nitrogen it needs by hunting insects.
- A carnivorous plant found can be found in the forests of Asia that can eat birds and even mice! Usually the animals mistakenly absorb the nectar of this flower and then get trapped in its chemical jars. The plant dissolves them in itself and makes them its food by producing digestive enzymes and absorbing nutrients.

- Plants have feelings the same as humans! Of course, they may not be able to express their feelings normally. But research shows that plants exposed to soft music will grow faster.

- Some types of bamboo grow up to 91 cm per day.
- The most poisonous plant in the world is Ricinus communis. Just 70 micrograms of this

pant are enough for an adult to lose his life. The venom in this plant is 12,000 times more deadly than the venom of the rattlesnake.
- There are some types of plants and animals that have evolved to live in strange places, such as inside volcanic pores, in huge caves, and even in the deep sea.
- A plant called the Larrea tridentate, some of which can live up to 10,000 years, grows in North America,
- Cyanide is a poison made from several plants. A small amount of it can be fatal in five minutes. In some countries, people sentenced to death are executed in a gas chamber with cyanide. In fact, they are locked in a chair in a gas chamber, and the executioner contaminates the space with gas from outside.
- In 1982, Japanese scientists found a 10,000-year-old plant seed in a pit! It is interesting to note that they succeeded in growing a tree after planting this seed!
- The strychnine tree has fruits similar to small oranges. But they are highly toxic. A small bite from the fruit of this plant can kill us!
- Mosses can live almost anywhere! They grow in cracks between small rocks that are only slightly exposed to sunlight.
- It may seem unbelievable, but we have had torrential rains throughout history in which fish,

frogs and toads have descended from the sky to the ground!
- Only 0.001% of the Earth's water is in the clouds and rain in one moment.
- In 1986, 92 people died in Bangladesh when hailstones, weighing more than 1 kg each, hit them.
- Heavy raindrops that landed in Brazil in 1995 were up to 8.6 mm in size. These droplets were measured by a laser.
- Raindrops fall at a speed of 11 kilometers per hour.
- Old stories and anecdotes about blood rains can be justified with red sands. They are transported by clouds over long distances and fall to the ground when it rains.
- Meghalaya is one of the northeastern states of India. It rains 1,187 mm annually there, making it the rainiest place in the world.
- Some clouds are 20,000 meters thick, which is almost three times the height of Mount Everest!

- It has not rained in parts of the Atacama Desert in Chile for 400 years.
- Rainforests cover only 2% of the Earth's surface, yet half of all plant and animal species on our planet live there.
- There may be about 200 different types of trees in a section of rainforest region about the size of a football stadium.
- Scientists estimate that millions of species of plants, insects, and organisms in rainforests have not yet been scientifically classified!
- In just one tree in the Amazon rainforest, 1,500 species of insects, including 50 species of ants, can be found!
- The world's biggest rainforest, the Amazon, is called the "lungs of the earth." It is a large forest in the Americas and covers an area about 5.5 square kilometers.
- The Amazon rainforest is not located in one specific country and is divided within 9 countries: Brazil (60%), Peru (13%), Colombia (9%), Venezuela (5%), Bolivia (5%), Guyana (3 %), Suriname (2 %), Ecuador (1.5 %) and French Guiana (1.5 %).
- The Amazon River, which flows through this forest, has more than 1,100 sub-tributaries! The river originates in the mountains of South America and flows into the Atlantic Ocean via a winding path.

- The importance of the Amazon rainforest is primarily due to the biodiversity of plant and animal species within it, and secondly because of its vital role in absorbing carbon dioxide and producing oxygen instead.
- Humans have always considered various aspects of wildlife on Earth as a potential threat to themselves, but the main threat to the planet Earth is man himself. In the last half century alone, the Amazon rainforest has shrunk by almost 20 percent!
- Some trees communicate using chemicals. If a *wood-eating* insect attacks such a tree, the tree releases chemicals into the air to repel the insect, which also triggers other trees in the area to produce the toxin to repel the insects.
- Trees grow from their tips. So if you engrave your name on a tree as a child (which you certainly do not), when you grow up, your artwork stays right there!

- Each normal tree produces an average of about 120 kilograms oxygen. This means that two mature trees simply provide the oxygen consumption for a family with 4 members for one year. It is unfortunate that between 2000 and 2005 alone, 80,467 square kilometers of the world's rainforest were destroyed by deforestation.
- Pine is an evergreen tree. The branches of these trees are never bare. Due to this feature, pine trees are widely used in municipalities and organizations. Even companies are interested in planting this tree in their environments and yards. It may seem unbelievable, but the roots of some pine trees can extend up to 48 km!
- The spruce is a tree in northern regions and its height reaches 20 to 60 meters.
- A Norwegian spruce with age of 9,550 years old has been found in the mountains of western Sweden. This tree has been called "Old Tjikko" and is said to be the oldest tree on earth.
- Recycling 1 ton of paper will prevent the cutting of 17 large trees (it is also possible to produce paper from stone and we may never have to cut a tree again).

- There is blood flowing in the human body, and when we intentionally or accidentally cut off part of our skin, that spot will bleed. But do trees have blood like us? The Australian strawberry tree has red sap and when we cut some part of this tree, something like blood flows from it!
- There is a native tree in South Africa called the quiver tree that automatically cuts off its branches to conserve water! Also, due to the shape of the branches, the natives of the region shave the hollow branches of this tree and use it as an arrow for their spears and for this reason, this tree is called the quiver tree.
- The smallest tree in the world is the dwarf willow tree that grows in Greenland. This tree is only 5 cm long!
- Some trees can live for thousands of years! The Giant Sequoia, for example, is a coniferous tree that is found more than anywhere else in the US state of California. The oldest of these trees is located in the Sequoia National Forest and is

3,000 years old (One Thousand year later Jesus Christ was born!).
- The Abarkooh Ancient Cypress Tree is one of the oldest trees in the world, which is at least more than 4500 years old and its planting is attributed to Zarathustra (who founded what is now known as Zoroastrianism).
- Parrotia persica, is a tree native to the northern forests of Iran. The wood of this tree is so strong and hard that it is known as chu-ahan-dar (meaning ironwood tree) among people living there.
- The bark of the Giant sequoia does not burn. However, the inside of this tree does not have such a feature and its inside will burn in case of fire.
- Many of the earth's Giant Sequoia trees were cut down by immigrants in the 19th century. Fire is also another danger element that threatens especially tall and old trees today.
- Trees like cypress and oak have more leaves and also produce more oxygen. Trees also produce more oxygen when they reach maturity. On average, one tree produces 114 kg of oxygen per year!
- There are trees that are more than 100 meters long! Giant Sequoia trees are also among the largest trees in the world. Some of them reach 180 meters in length.

- Lack of light affects the appearance of the tree and its speed of longitudinal growth. Trees that grow in the light usually have thicker trunks, depending on the species..
- When a tree grows, new layers of wood form under its bark. The color and shape of the layer that is made in summer is different from the color of the layer that is made during winter. Therefore, the presence of a lighter layer next to a bolder layer indicates that one year of the tree lifespan has passed. After cutting the main trunk of the tree, its age can be reliably estimated by counting the dark and light rings of the tree

trunk. If a tree has 5 rings on its trunk, it is almost certain that it is a 5-year-old tree.

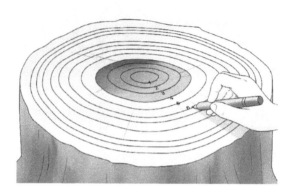

- In the most modern way to estimate the age of a tree, they make a thin hole in the trunk of the tree to its center. Then a special liquid is poured into the hole. This liquid will harden after a while. When it is removed from the trunk of the tree, the engraving of layers (described above) remains on it, and by counting these layers, the age of the tree will be measured.
- A phenomenon in which the tops of tree branches avoid contact with each other is called a crown shyness. Some believe that this indicates a kind of privacy between tress.
- Although tomatoes are considered a fruit in botany, they are often referred to as vegetables in agriculture due to their lack of pits.
- Tomatoes were firstly called Roman eggplant in Afghanistan and Armenian eggplant in Iran

because they were imported from western countries.
- In a book published in England in 1597, it was claimed that tomatoes were poisonous! At the same time, the author acknowledged that the fruit is regularly eaten in Spain and Italy, despite being toxic. In fact, tomatoes contain small amounts of taumatin, a type of glycoalkaloid, a toxin that is usually not dangerous. The author's views influenced public opinion, and the plant was recognized as an unhealthy food in Britain and its Colonies in North America and was not used until the 19th century.
- Pumpkin roots can be up to 24 km long!
- There are more than 38,000 species of mushroom, some of which are poisonous and some of which are edible. The color, shape and size of the mushrooms can be very varied.
- The largest living thing on earth is a giant mushroom that covers ten square kilometers! This fungus is located in Oregon, USA and is thought to be between 2400 to 8650 years old!
- There is a type of fungus in Africa that grows faster than any living thing in the world. This plant grows to 20 cm in just 20 minutes. You can hear it flourish as it grows!
- Some fungi glow in the dark and can be seen from a distance of 15 meters. They can be used as natural lanterns!

- The potato belongs to the belladonna plant family. Ironically, the belladonna plant known as deadly nightshade is considered the most toxic plant in the Western Hemisphere. If potatoes had been discovered now, they would probably not have been approved as a food and would have initially been viewed with suspicion!
- Potatoes were transported from South America to Europe in 1570 by the Spanish. The fruit was later brought to North America by British colonists.
- The potato was first brought to Iran by John Malcolm, British military officer in the middle of Fath Ali Shah Qajar era, which is why it was originally called the "Malcolm plum"!
- Potatoes are called the "apple of Earth" in some languages. They are no more fattening than a

normal apple! They have about the same amount of calories. It is the way potatoes are cooked (often with oil or butter) that makes them fattening.
- Extinction is a condition in biology and ecology in which the last member of a species dies and the other species can no longer survive by natural reproduction. Every year, 140,000 species of animals become extinct. Some researchers believe that half of the world's current species will be extinct by 2100.
- Some scientists estimate that more than two million species of animals became extinct during the twentieth century alone.
- Many things have happened on Earth of which only traces and remnants left for us. Researchers believe that at least five events have occurred in the last 550 million years, each ending the life of at least half of the planet's living things.

- A mirage is a picture of something that does not really exist. This phenomenon is caused by the impact of a layer of hot air on a layer of cold air.
- On July 7, 1987, the temperature in Kansas USA rose from 24 degrees Celsius to 35 degrees Celsius in 7 minutes.
- The phenomenon of the appearance of colored and moving lights in the night sky is called aurora and usually occurs in latitudes close to the two poles of the earth.
- Auroras are also called polar lights because they are often seen in the Northern Hemisphere, and the closer you are to the North Pole, the more likely you are to see them due to their proximity to the Earth's magnetic pole. They can be seen in cities in northern Canada that are very close to the poles.
- The largest recorded iceberg was found on November 12, 1956, west of Scott Island in the South Pacific. This iceberg, belonged to Antarctica and had an area of 32,300 square kilometers (334 kilometers long and 96 kilometers wide), which is larger than Belgium!

- In 1849, a large piece of ice (6 meters long) fell from the sky in Scotland!
- Naturally, a liquid such as water will freeze when exposed to extreme coldness, but it is interesting to note that hot water freezes faster than cold water! This phenomenon is called the Mpemba effect.
- The strongest winds blow in the Gulf of Commonwealth in Antarctica, with winds of about 322 kilometers per hour.
- The coldest places on Earth are the North and South Poles. The reason for this is the lack of regular sunshine impact with them. There is no sun in Antarctica for 182 days of the year! This is 186 days in the Arctic.
- Hawaii is an archipelago in the Pacific Ocean and is the 50th state in the United States. There is a green sandy beach on a Hawaiian island

called Papakōlea Beach! This is due to the presence of a mineral called olivine, which separates from volcanic rocks and crumbles on the shore.
- Scientists hope to genetically engineer a deep-sea fish gene into a vegetable to make it cold-resistant! Some people call genetically modified foods, Frankenstein foods!
- There is a river in Colombia that has five colors. The name of this river is Caño Cristales. Due to mineral stones, the riverbeds are filled with algae that have different colors including blue, green, yellow, and purple.
- Some cold-water coral reefs have been growing since the end of the last ice age (10,000 years ago).
- Fire moves faster uphill than downhill! Because the hot air from the fire moves upwards, it heats and dries the trees that are higher, so the burning operation is easier that way.
- The coldest place on earth is near Russian Vostok Station in inland Princess Elizabeth Land, Antarctica. Founded by the Soviet Union in 1957, the station has recorded lowest reliably measured natural temperature on Earth of -89.2 °C.

- There are as many as 5 billion bacteria in a fingertip! If we want to divide this number of bacteria, there are about two bacteria per person in Asia! Of course, not all bacteria are so tiny. The largest bacterium discovered is one millimeter long. That is, it is large enough that you can see it with the naked eye.
- Many creatures have unique characteristics that adapt to the environment where they live. In the Namib Desert Africa, for example, there are plants called "windows" that form clear crystals on their leaves to protect themselves from the hot sun.
- In 90% of all cases, an avalanche starts due to a movement by humans.

- When you are stuck in an avalanche, if you are saved in the first 15 minutes, you have a 93% chance of survival, if you stay under the snow for 45 minutes, this chance is between 20 and 30%. It is very rare for someone to survive in the snow for more than 2 hours.
- The frequency of storms and tornadoes on Earth's atmosphere has doubled in the last 100 years (probably due to human intervention in nature)!

Inventions

- The wheel can undoubtedly be considered as one of the most important inventions of humanity. The credit for the invention of the wheel is attributed to the Elam civilization because their sculptures are the oldest surviving evidence of a wheel. Also in Mesopotamia, images of wheeled wagons, dating from between 3500-3350 BC, were found on clay screens attributed to Sumerian civilization.

- The invention of the printing press had an incalculable effect on the intellectual progress of humanity. There is a general perception that Johann Gutenberg formed the printing industry in 1456, but what he actually did was invent the printing press. Gutenberg built a printing press in his workshop using technologies and equipment invented before him.

- Gutenberg's machine used a separate piece of metal for each letter, and these pieces were put together to create words. The above letters were then ground on ink and pressed onto paper. Gutenberg first made letter pieces from wood, then from lead, and later from an alloy of lead, tin, and antimony.
- The world's first car was built by a French engineer named Nicolas-Joseph Cugnot. In 1770, after seven years, he succeeded in building the world's first all-mechanical car. The car was in the shape of a tricycle and had a large boiler in front of it.

- Cugnot built the device with the intention of helping the French army, and the car traveled only 3.6 kilometers per hour. However, using a steam engine for this purpose opened a new path for other inventors!
- English mechanical engineer George Stevenson was the first to use steam power to propel a locomotive. The importance of Stevenson's work was that he started from almost zero and tried everything from being a miner to a shepherd to earn for a living. Stevenson spent 15 years trying to build a locomotive and finally succeeded.
- Stevenson launched the first train from Stockton to Darlington in 1825, and it has since become the most popular source of transportation for more than a century. George Stevenson's son, Robert, also followed in his father's footsteps.

- One of the most important inventions of the eighteenth century was the camera. This invention revolutionized art and is known as the spark for the creation of visual media. But who is the inventor of the camera? The oldest surviving photograph was taken by Nicéphore Niépce about 200 years ago, and many credit him as the inventor of photography.

- In the image left from the Niépce collection, sunlight shines on the buildings on both sides, which means that the photography lasted about 8 hours. Using chemical compounds and a metal plate, Niépce was able to capture a view from outside his room.

- According to Europeans, the first camera was invented by Newton. Some attribute the invention to Robert Hooke.
- Nowadays, while cell phones have pushed the boundaries of the world, we cannot talk about the inventors and explorers, without mentioning

inventor of the telephone. The Scottish-American inventor Alexander Graham Bell is known as the inventor of the telephone. Before him, several other scientists were working on electronic voice transmission and were in fact a kind of competitor, but eventually Alexander Graham Bell was the first to patent the electric telephone in 1876. His other famous inventions include the metal detector, the gramophone, and the twisted pair wire.

- Thomas Edison, an American engineer and inventor, completed or invented a variety of devices. Edison's most important invention is the incandescent light bulb. There have been criticisms of Edison's performance. Some believe that he was not actually the inventor of

the lamp, but someone named Humphrey Dewey did it, and that Edison succeeded in mass-producing his invention.
- On the night of December 31, 1879, Edison invited journalists and investors to Menlo Park in New Jersey to demonstrate his new invention. He ordered the laboratory to be lit with hundreds of light bulbs to create a light-filled view. This light in the darkness of the night surprised the guests so much that Edison soon found many supporters and investors.

- The first mechanical watch was made in Europe in the 13th century. However, none of those early clocks are left at present, and we know that they existed only because they are mentioned in the records and documents of medieval churches.

- The most accurate clock in the world is the atomic clock and is used as an international standard for scheduling.
- The oldest type of clock used in Iran is the water clock. In these clocks, water flows with a certain intensity from container A into container B. Callisthenes, a Greek historian in 300 BC, mentions the use of this type of clock in Iran.

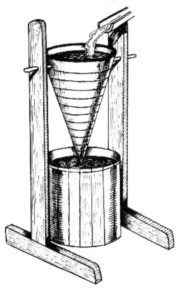

- The first car equipped with a gasoline or diesel engine was built in 1885 by a German inventor named Karl Friedrich Benz. The car was powered by three wheels and a four-stroke single-cylinder engine with 0.69 hp at a maximum speed of 16 km/h.

- Three years after the invention of the gasoline car, Benz's wife drove for the first time, a 104-kilometer journey from Mannheim to her hometown of Pforzheim in 12 hours and 57 minutes. In 1893, Benz also invented his first four-wheel drive gasoline car.

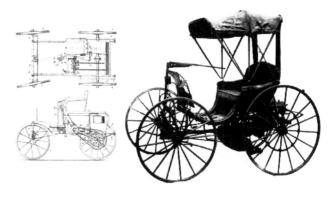

- The Wright brothers, Orwell and Wilbur, were two Americans from Ohio best known for inventing the first heavier-than-air controllable aircraft on December 17, 1903.
- The Wright brothers worked on gliders and vehicles. Of course, long ago other people were able to build flying machines and even fly with them, but the Wright brothers' invention was the first flying device that was completely controllable by pilot.

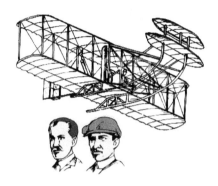

- The radio was invented by Nikola Tesla, the Serbian inventor, electrical engineer and mathematician. However, there is a general belief that radio was invented by Guillermo Marconi. For decades the invention of radio was registered under the name Marconi because he was the one who developed the public use of it. But in 1943, the US Supreme Court overturned Marconi's patent and granted it to Tesla. Ironically, both had died at this time. Some believe that the support of Thomas Edison and Andre Carnegie for Marconi is the main reason why Tesla was denied the benefits of his invention in its early years.

- The first missiles were used as weapons of war by Nazi Germany in World War II. Wernher von Braun, a German scientist and member of Schutzstaffel, is considered the pioneer and father of the missile industry in the world. The missiles made a huge difference in all aspects of military operations, especially air and sea battles.
- These rockets formed the basis of space researches after World War II and with the founding of NASA, von Braun, who had no choice but to cooperate with the victorious side of the war played a key role in determining US space goals and successes.

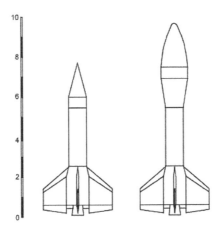

- The Persian equivalent of television is the word "Doornama" (means far view) and "television" is the French word itself. A Scottish engineer named John Logie Baird first built a television set in 1925 and succeeded in transmitting motion pictures from one place in London to another point. Although the original images were of poor quality, this was a great discovery for humanity!

- Credit cards are a type of payment card through which individuals or legal entities can conduct their financial transactions. Credit cards were first used in the United States in the 1950s. A company called Diners agreed with several restaurants to offer cashless service between them. Customers could use the restaurant by presenting a Diners card and pay Diners the money at the end of the month!
- The World Wide Web should be considered the largest and most complex system ever designed, engineered and implemented by humans. This huge global network was built by the Pentagon in the 1960s with the goal of building strong communications networks so that communications would not be cut off even in the event of a nuclear attack. However, it was not until the 1990s that the Internet became a public network.
- Ed Roberts was an American physician and engineer who invented the first personal computer in 1975 and is known as the "father of personal computers." Roberts founded MITS in 1970, and his company's electronic calculator kit quickly gained attention, and in November 1971 he appeared on the cover of Popular Electronics magazine.

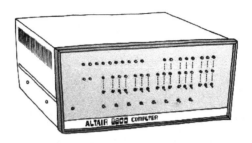

- Roberts developed the Altair 8800 personal computer in the mid-1970s, which used the new Intel 8080 microprocessor. It was at this time that Bill Gates and Paul Allen joined his company and Microsoft emerged from their efforts. Interestingly, Roberts left the industry entirely in 1977, despite a bright future for the computer industry.

Images

The Oxford Dictionary defines science as "knowledge about the structure and behavior of the natural and physical world, based on facts that you can prove, for example, by experiments."

Science encompasses a variety of branches, from chemistry to physics and from biology to astronomy. Humans owe much to the great scientists of history and their great achievements. This chapter features images of some of history's great scientists

1. Archimedes

2. Isaac Newton

3. Leonardo da Vinci

4. Galileo Galilei

5. Michael Faraday

6. Charles Darwin

7. Louis Pasteur

8. Nikola Tesla

9. Marie Curie

10. Albert Einstein